Dinosaurs, Diamonds, and Things from Outer Space
The Great Extinction

David Brez Carlisle

Dinosaurs, Diamonds, and Things from Outer Space

The Great Extinction

STANFORD UNIVERSITY PRESS

Stanford, California • 1995

Stanford University Press, Stanford, California

© 1995 by the Board of Trustees of the Leland Stanford Junior University

Printed in the United States of America

CIP data appear at the end of the book

Stanford University Press publications are distributed exclusively by Stanford University Press within the United States, Canada, and Mexico; they are distributed exclusively by Cambridge University Press throughout the rest of the world.

For Roxane,
who shares my joy in science and in music, and an
insatiable curiosity about the universe,
without whose support I should never have
attempted this book.

Preface

The idea that an object from outer space—the general name is a bolide—hit the earth 65 million years ago has been around for more than a dozen years. We may suppose that this bolide wiped out the dinosaurs, pterodactyls, and ichthyosaurs, completely stopped the formation of chalk, and clobbered life generally.

Since 1980 we have known of evidence that this actually happened, that 85 percent of the then-existing species of animals were killed off at this time as the result of some kind of object from outer space. But what kind of object? Was it a fragment of an asteroid? Or perhaps a comet? Maybe it was an Apollo body or a chunk of rock thrown off the moon. And what of other mass extinctions? Did they result from the same mechanism? Were they all part of a series that repeats at regular periodic intervals, or were they somehow independent, different from one another, unrelated? And where do supernovas come into all this? Or diamonds?

In this book I detail my hunt for geochemical evidence as to the precise nature of this Cretaceous-Tertiary Boundary event of 65 million years ago, and I produce a scenario that succeeds for the first time in explaining the reasons why certain groups of animals were extinguished while others survived, why lizards lived on while their cousins, the "terrible lizards," were wiped out.

In carrying out this research, I have found it necessary to develop a new model of the origin of the solar system and, for the first time, offer a coherent account of the formation and place of the comets. Though the basis of this model involves rather deep mathematics, I explain it here in words and pictures, without ever offering a mathematical formula.

Finally, I have tried to give some flavor of the excitement of doing science, the fun of making new discoveries, whether through theory or observation. And I am completely biased toward my own theories. If you want to read about what Joe Bloggs thinks of all this, read his works for yourself; don't expect an unbiased account of what he has said from me. I indulge in controversy to the hilt, revel in it, and expect other people to try to put me down. Of course I stand on the shoulders of my predecessors in this field, but what fun to be able to convince myself that I can see farther than they!

D.B.C.

Acknowledgments

I have benefited enormously from discussions with numerous authorities, and I should like in particular to thank (in alphabetical order) the following: B. K. Afghan, National Water Research Institute, Burlington, Ontario; Walter Alvarez, University of California, Berkeley; Frank Asaro, University of California, Berkeley; Dennis R. Braman, Royal Tyrrell Museum of Palaeontology, Alberta; John Cronin, Arizona State University; Richard A. F. Grieve, Geological Survey of Canada; Martin J. Head, University of Toronto; W. E. Keiser, IsoTope Laboratory, University of Toronto; Linas Kilius, IsoTope Laboratory, University of Toronto; A. E. Litherland, IsoTope Laboratory, University of Toronto; Digby MacLaren, Royal Society of Canada; Ray McLenaghan, Canadian Institute for Theoretical Astrophysics; James J. Pospichal, Florida State University; John C. Rucklidge, University of Toronto; Pearl Weinberger, University of Ottawa; John Westgate, University of Toronto; Zhao Xiao-Lei, IsoTope Laboratory, University of Toronto.

John Westgate and Barbara MacLean, of the Geology Department, University of Toronto, facilitated all arrangements for me during the writing of this book. Bill Glen of the U.S. Geological Survey served as my scientific editor and was a constant inspiration, and Ellen F. Smith of Stanford University Press and Evelyn Mercer Ward did a superb job of editing the manuscript. The late Pearl Weinberger, University of Ottawa, was a constant friend and helped me in my early laboratory work for this project before her untimely death. Karyn Gorra and Philip Ower took some of the photographs. Gary Thomas helped me with the graphics.

Since this book is intended for the general reader, and not only for the specialist, I pestered some of my friends who were not professional scientists into reading parts of the manuscript at various stages of development. I should particularly like to thank Carol King, Robert and Celia Kornfeld, Ambassador Jan Nadelman, and Baroness Gabrielle von Neumann de Vegvar for their patience and encouragement.

Richard G. Langdon, director of the Thomas Fisher Rare Book Library of the University of Toronto and members of his staff were unfailingly helpful and courteous. The assistant director, who also curated the Memorial Exhibition for the centennial of the death of J. B. Tyrrell, kindly provided two of Tyrrell's original photographs used here, including the solitary picture of a dinosaur fossil.

Contents

Tables

Figures

A Note on Jargon

Even so familiar a word as *animal* has different meanings in science and in daily speech. For myself, I always use this word in its scientific sense to mean a member of the animal kingdom, whether it be bird or beast, insect or *Amoeba*—all are animals to me, which means I am using jargon when I write *animal*. Yet in common speech, more often than not, *animal* means a furry creature that suckles its young, what a scientist would call a mammal.

Sometimes, though, in colloquial speech *animal* means anything with a backbone, fish or fowl, mammal or reptile. And then there are other disciplines where *animal* has yet other meanings. Did you know that according to Canadian law beaver and fox are not animals? Or that a cat is not furry? In legal jargon in Canada beaver and fox are "furbearers," but the Canadian Environmental Protection Act has defined *animal* to exclude all furbearers. The way these terms are defined means that a cat is an animal but not a furbearer, while beaver and mink are furbearers, not animals.

So jargon is not a prerogative just of scientists. That said, I shall try to avoid jargon in this book as far as I am able. Jargon, however, is inescapable. It can originate in more than one way.

The first way is illustrated above: the redefinition of a common word in order to avoid ambiguity. Often enough common usage and specialized usage of the word continue side by side, as with *animal*, and in a scientific (or legal) context jargon usage must be faced. I make no apology for using *animal* in its strict scientific sense, but there are a few other terms that require some explanation. For instance, epoch, era, and period all have specific meanings in geology that are more restricted than common usage. I shall try to avoid other words with specialized meanings, but they should not make life too difficult if a few do slip in.

Sometimes the common usage of a word has disappeared entirely. An educated man during the eighteenth-century Enlightenment (and I do mean man, not person, for during the Enlightenment women's education was discouraged, universities would not admit women, and women had no vote) would know that the word *cretaceous* referred to chalk cliffs and deposits, for he had been taught Latin at school and the Latin for chalk is *creta*. Today the word has dropped out of use entirely except for its geological meaning for the Cretaceous period, which terminated 65 million years ago.

A second way that jargon develops is as a form of shorthand for complexities, as abbreviations almost, sometimes acronyms. I have used as few of these as possible and explained them (not defined them) in the glossary at the end. Quite often there is no real excuse for this type of jargon. Why do we have to write "orogenesis" when "mountain building" will usually do just as well? Some of us do love polysyllabic words! Some ideas are more complex, though, and it becomes too cumbersome to write about them without using some kind of shorthand jargon; some are both complex and of peripheral interest only, so it is not worthwhile to explain them at length. Such analytical tools as accelerator mass spectrometry (AMS) and inductively coupled argon plasma analyzers (ICAP) receive brief mention in this book (and are briefly explained in the Glossary), but since they are only tools for my purposes I have not attempted to write about them in any fully comprehensible way; I have just allowed myself to sink into jargon in these instances. Failure to explain such tools as AMS more fully is not due to ignorance of the topic (I edited a book on AMS published by the Natural Sciences and Engineering Research Council of Canada in 1991), but rather to my desire to keep the focus on the real topic of this book.

Jargon may develop in science by a third route, that of naming a newly discovered phenomenon or object. These names are necessary, and the first time I use them I also include a lengthy explanation. Such terms are often the most difficult jargon to understand because the very concepts are unfamiliar. Sometimes, however, these words penetrate so fully into the public consciousness that they cease to be jargon, despite the initial complexities of the concept. Take the word *dinosaur*, for instance, a word so familiar to every child (at least in the "developed" world) that it can no longer be regarded as scientific jargon. Yet it is only 150 years since the word was first deliberately coined by Richard Owen (1841). A more recent example is DNA, part of common vocabulary now, a mere acronym standing for deoxyribonucleic acid. Incidentally, I can find both *dinosaur* and *deoxyribonucleic* in the dictionary of my word processor, yet *spectrometry* is absent.

That's enough about jargon. I will avoid it when I can, but beware, some is unavoidable! Use the glossary if you have to; it gives meanings, not definitions. And if you don't know what *taxon* means, look it up.

. . . .

Dinosaurs, Diamonds, and Things from Outer Space
The Great Extinction

Oh, Diamond! Diamond!
You little know what Mischief is done!
—Sir Isaac Newton

1

Introduction

This is the way the world ends.
This is the way the world ends.
This is the way the world ends,
Not with a bang but a whimper.
—T. S. Eliot, *The Hollow Men*

What nonsense! Here is one of the best-known quatrains of twentieth-century poetry, yet such nonsense, such utter nonsense!

If a present-day poet were to write those lines we should immediately conclude he was fearing imminent death. They are a heartfelt cry of despair for the end of a personal world. Yet this kind of whimpering end of life and of one's personal world is hardly the norm, even in the developed world. The world of American male teenagers more often ends in a bang: gunshots are now, in the United States, the leading cause of death of male teenagers—both black and white—and urban blacks have less than 50 percent chance of reaching the age of 21. Their world does not end with a whimper, but in the bang of a gun.

Throughout history, violent death has been the norm. The killing fields of Cambodia, the bombardment of Sarajevo, genocide in the Sudan, tribal warfare in Sierra Leone, religious strife in India—these are the norm, not the often-sanitized Western hospital death, with bodily functions prolonged beyond all reason. An ailing antelope has no hospital; it cannot just curl up in a corner and die peacefully. Instead it will suffer a violent death from hyenas, Cape hunting dogs, lions, or even vultures.

Species, too, more often than not, come to an end with a bang. Let me digress for a moment to the national dish of Quebec, *la tourtière*, a sort of meat pie, made nowadays from a mix of ground veal, pork, turkey, and other meats and spices. Delicious! But its name betrays its origins: a pie made from the meat of *tourterelles*, which in France means turtledoves or pigeons but in Canada meant passenger pigeons. The great flocks of passenger pigeons that darkened the skies of nineteenth-century North America fell to the shotgun to fill the meat pies of Quebec and the stews of New York.

At first the pigeons were a strictly local delicacy. In 1814 John James Audubon witnessed a passenger pigeon hunt in Ohio:

As the time of the arrival of the passenger pigeons approached, their foes [the local farmers and hunters] anxiously prepared to receive them. Some persons were ready with iron pots containing sulfur, others with torches of pine knots; many had poles, the rest, guns. . . . Everything was ready and all eyes were fixed on the clear sky that could be glimpsed amid the tall tree-tops. . . . Suddenly a general cry burst forth, "Here they come!" The noise they made, even though still distant, reminded me of a hard gale at sea, passing through the rigging of a close-reefed vessel. The birds arrived and passed over me. I felt a current of air that surprised me. Thousands of pigeons were soon knocked down by the polemen, whilst more continued to pour in. The fires were lighted, then a magnificent, wonderful, almost terrifying sight presented itself. The pigeons, arriving by the thousands, alighted everywhere, one above the other, until solid masses were formed on the branches all around. Here and there the perches gave way with a crack under their weight, and fell to the ground, destroying hundreds of birds beneath. . . . The scene was one of uproar and confusion. . . . Even the gun reports were seldom heard, and I was made aware of the firing only by seeing the shooters reloading. . . .

The picking up of the dead and wounded birds was put off till morning. The pigeons were constantly coming and it was past midnight before I noticed any decrease in the number of those arriving. The uproar continued the whole night. . . .

Towards the approach of day, the noise somewhat subsided. Long before I could distinguish them plainly, the pigeons began to move off. . . . By sunrise all that were able to fly had disappeared. . . . Eagles and hawks, accompanied by a crowd of vultures, took their place and enjoyed their share of the spoils. Then the authors of all this devastation began to move among the dead, the dying and the mangled, picking up the pigeons and piling them in heaps. When each man had as many as he could dispose of, hogs were let loose to feed on the remainder.

As the railways began to extend across the continent, the hunting of passenger pigeons began to be more organized, and the trade to the cities became routine. Hundreds of tons of pigeon carcasses were sent to the great cities of New York and Montreal, Cincinnati and Chicago; pigeon pie and *tourtière* were no longer country cooking but city staples. By 1914 the last passenger pigeon had died in Cincinnati's zoo; the hunters' gun had wiped out a species.

The American bison nearly met the same fate to feed the troops in the

Indian wars, and the dodo of Mauritius was extinguished by seamen who clubbed the birds to death to fill their stewpots as a change from salt beef.

This book is primarily about the bang that terminated the Mesozoic Era and the mass extinction that resulted. It is a frankly biased account, giving my own interpretations of other people's observations as well as of my own. Where I present others' views, it is often as a counterfoil for my own, either to form a basis upon which to speculate or to contradict them. If the reader wants to learn about Joe Bloggs's views, then he must read Joe Bloggs's writings; after all, Joe can present his own views better than I can. I make no pretense to be impartial. I have had fun doing the science. I have had serious fun thinking about the science and the problems, arguing vigorously with other scientists about it all. I hope that in this book I am able to convey something of this fun, for science is above all fun. Forget the dreary image of the scientist as a drudge in a white coat or, worse yet, as a madman. A good scientist is usually getting more enjoyment out of his activities than a hockey player. After a lifetime in science I am only grateful that I have been able to find employers willing to pay me to do what I enjoy: the physical thrill of fieldwork, the puzzle solving at my desk or at the computer, the excitement of working at the bench (and usually discovering that my initial ideas were hopelessly wrong), and above all creating mental images worth testing in the lab. Doing science is having serious fun.

The key to understanding the extinction of the dinosaurs—and of all the other creatures that were wiped out at the same time—lies in understanding the slight clues—diamonds and amino acids, traces of metals and radioisotopes—that can be gleaned from the rocks laid down at that time, the time when the Mesozoic era ended and the Tertiary, or Cenozoic, era began. In the rocks this is generally known as the Cretaceous-Tertiary Boundary, or K-T Boundary for short. (K comes from the German name for Cretaceous, so as not to mix it up with the Cambrian or Carboniferous periods.) Terminologically, this is a complete cock-up, mixing different kinds of names, but that is the term geologists have adopted.

The clues are not found by accident, of course. Each slight anomaly uncovered leads to intensive theorizing. Each new theory propounded, if it is to be at all useful, must make predictions that can be tested. A hypothesis in science is valueless unless it makes predictions; otherwise it is a mere scenario without substance. Too many scenarios have been written about the Cretaceous-Tertiary Boundary that do not make testable predictions. Of course a good hypothesis should be presented also as a scenario to help others understand it, and I shall attempt to do so in Chapter 13, but the development of the underlying theory and the steps taken to test it are the bread and butter of science. Detective work lies in interpreting the clues and in developing theories that may uncover other clues. Sometimes the physical resources that must be mobilized to test an idea are so formidable that it is not possible to put them together without a strong theoretical underpinning to show it is worthwhile. The use of multimillion-volt accelerator mass spectrometry is one example of this.

Accordingly, in this book I shall intertwine the unraveling of physico-chemical clues with theory—and attempt to explain the latter without resorting to the mathematical equations with which I have struggled in developing the hypotheses. I shall run the gamut from nuts of cycad trees to the origin of the solar system, from the formation of plutonium in supernovae to the egg laying of the pearly *Nautilus*. All of these are significant when we come to consider just what did happen at the end of the Cretaceous period.

I shall also try to resolve the disputes between the various scenarios that have been advanced for the Cretaceous-Tertiary event—an event marked by a distinct stratum in the rocks and indicating the moment when not merely the dinosaurs were killed but also 85 percent of all species. This event has been variously ascribed to vulcanism or to the impact of an extraterrestrial body. I don't believe a word of the vulcanism theory. The extraterrestrial body may have been either an asteroid or a comet. I shall present detailed evidence that it was a comet or, more likely, a shower of comets. Four scenarios have been advanced recently giving reasons why the comets could have been perturbed from their accustomed courses. I have written down the equations governing these scenarios, and I will show that none of them is at all feasible. I shall not present the detailed mathematics, but instead explain in words what it is all about. Finally, I shall present my own scenario, a scenario involving a nearby supernova, steam-propelled comets, diamonds from outer space, and acid rain.

This scenario evolved through an iterative process of theorizing and experimentation. Each new stage of theorizing led to a search for new experimental evidence, and every morsel of experimental evidence led to new theories. The important thing about any theory is that it should lead to the possibility of experimental disproof. A theory that makes no predictions is useless.

Life seems to have begun on earth about 3.8 billion years ago, at least life as we know it. It seems likely that life of some kind evolved at least once before that date in the half billion years since the formation of the earth, but was wiped out by the bombardment of the earth by planetesimals, the objects that created all the giant craters on the moon. But that is pure speculation. Few rocks have survived dating from before 3.8 billion years, and those that have show no trace of fossils of any kind. This era, from the creation of the world until the earliest traces of life, is known as

TABLE I.I
A Brief Table of Geological Time

Era	Period/system	Age (millions of years)
	Neogene	
	Quaternary (Recent)	0.01
	Pleistocene	1.6
	Pliocene	5
Cenozoic	Miocene	24
(Tertiary)	Paleogene	
	Oligocene	37
	Eocene	58
	Paleocene	65
	Cretaceous	144
Mesozoic	Jurassic	208
(Secondary)	Triassic	245
	Permian	286
	Carboniferous	360
Paleozoic	Devonian	408
(Primary)	Silurian	438
	Ordovician	505
	Cambrian	590
	Precambrian	2,500
Proterozoic	Archaean	3,800
Azoic	Hyadean	

the Azoic era, meaning the era without life. This is followed by the Proterozoic era, meaning the era of the first animals, extending from 3.8 billion years to about 590 million years ago. In the earlier part of this long stretch of time, the few fossils that have ever been discovered are all of single-celled organisms, presumably primitive bacteria of some kind, chiefly photosynthetic organisms that can, like plants, use the energy of sunlight to synthesize their nourishment. This period is known as the Archaean period, the earliest subdivision of the Proterozoic era (Table 1.1).

In the succeeding Precambrian period multicellular creatures evolved, but the creatures that existed had no hard parts, so fossils are extremely rare. Only at a few localities, such as the Ediacara Range in Australia, were conditions right for soft-bodied animals and plants to be fossilized.

Then came the Cambrian "explosion," when many groups of animals developed hard parts that could readily be fossilized. It seems that the Proterozoic era ended in a bang, probably an impact from outer space, and that the few survivors began, in the Cambrian, to compete with hard jaws to crush their prey and then with armor for protection against hard jaws. This explosion of species formation, 590 million years ago, ushered in the Paleozoic era, the first era from which we find abundant fossils. The oldest period of the Paleozoic is the Cambrian period. By no means all, or even the majority, of the animals of this period had hard parts, but it is the fossils of hard parts that are best known. The discovery of the Burgess

Shale, a fossil bed near Trail in the Rocky Mountains of British Columbia, in which the unusual conditions of preservation had led to the fossilization of soft-bodied animals, made us realize just how many species there were that did not have any hard parts at all. As if we did not know this already from observing living animals! And the hard parts we do know well from Cambrian fossils are often impossible to relate to any animals we know today. Trilobites are obviously relatives of modern crustaceans, but conodonts, for example, were first thought to be teeth, then some kind of internal support for some kind of organ, perhaps related to worms, perhaps to vertebrates, perhaps to some creature we do not know at all. I happen to believe that they represent stiff supports for the feeding apparatus of a creature distantly related to the present-day amphioxus.

Like the Proterozoic era the Paleozoic ended with a bang, about 245 million years ago. About 95 percent of all fossil species disappeared at this moment in time, a really major extinction, to be succeeded by the Mesozoic era, the so-called Age of Reptiles (Figure 1.1). The few surviving individuals and species diversified in new directions, just as with the Cambrian explosion, leading to the multiplicity of reptiles and of mammals.

The Mesozoic era too ended in a bang 65 million years ago (which is the main subject of this book), wiping out a mere 85 percent of species, and was succeeded by the Cenozoic era, the Age of Mammals, which continues to the present day.

The four main fossiliferous eras are, in succession, the Proterozoic, Paleozoic, Mesozoic, and Cenozoic, meaning respectively the eras of the "first" animals, the "old" animals, the "middle" animals, and the "modern" animals. The last three are grouped together into the Phanerozoic, meaning the era of readily visible fossil animals. The Proterozoic, Paleozoic, and Mesozoic each ended in a bang, with mass extinctions, as indeed did the preceding Azoic era. We may suppose that the Cenozoic will end in the same way. The three eras of the Phanerozoic are also called the Primary era (Paleozoic), Secondary era (Mesozoic), and Tertiary era (Cenozoic). It is sometimes convenient also to add a fictitious Quaternary era, the Age of Man, though there is little to distinguish it from the Tertiary. Each era ended with a cataclysm. What will happen to our own era? Will it end in a cometary disaster such as has recently been suggested as a possibility for a year early in the twenty-first century, when comet Swift-Tuttle next makes an apparition? Will this comet actually hit the earth and wipe out much of life here? Probably not. At least not so soon.

Modern geology, after a hesitant start, really took hold with Sir Charles Lyell, who was largely responsible for the general acceptance of the idea (originated by James Hutton) that geological processes have been uniform over time and that all the events we see in the rocks have occurred because

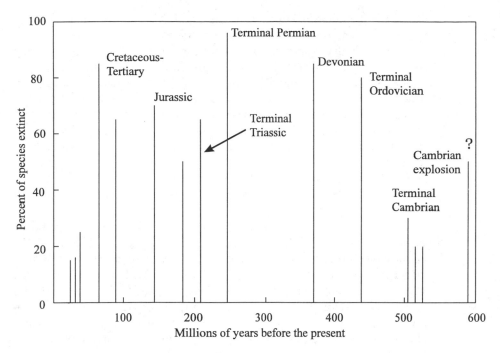

Fig. 1.1. Mass extinctions during the last 590 million years. A mass extinction is defined here as the extinction of at least 10 percent of known species in a period of a million years or less. The height of the spikes indicates the percentage of species that became extinct at each known mass extinction. The major extinctions are labeled. The Cambrian explosion, 590 million years ago, marks the origin of animals with hard parts. Because almost no animals left any fossil traces before this event, we have little information about its magnitude, but its consequences imply that it was a major extinction.

of processes that are still occurring today. Indeed, he proposed that the long periods of time needed for current natural processes to produce the effects we see in the rocks demanded that the earth be far older than the 6,000 years of the official church chronology and suggested that it might be as much as a million years old. His influential book on the *Principles of Geology*, which so influenced Charles Darwin, was published shortly before the voyage of HMS *Beagle*. Uniformitarianism, or gradualism, was firmly entrenched in scientific orthodoxy by the time *The Origin of Species* was published. Lyell finally, in his view, put the old idea of cataclysms (and of Noah's flood) to rest in 1858 when he surveyed Mount Etna in Sicily and showed conclusively that it had been built up by repeated small eruptions rather than by one gigantic cataclysmic upheaval.

Gradualism, under the influence of Lyell and Darwin, became palaeontological and evolutionary dogma and orthodoxy.

But the theory of cataclysmic change refused to lie down. The first rumblings of modern catastrophism came in the 1920s when the British zoologist Walter Garstang began to disseminate his ideas that vertebrates originated by sudden abrupt processes, which today we should call saltatory evolution. In the late 1940s and early 1950s I tried halfheartedly to follow up on this work in some of my own papers on protochordates but soon abandoned this line of inquiry (Figure 1.2). It was not until Stephen Jay Gould of Harvard University and his associates in the 1970s produced the theory they called "punctuated equilibrium" (everyone else calls it "punk rock") that it became acceptable to think of evolution proceeding stepwise, not uniformly.

Then in 1980 the Berkeley group of Luis and Walter Alvarez and their associates produced the seminal paper suggesting that the event at the Cretaceous-Tertiary Boundary that produced such massive extinctions was the result of the impact of an asteroid, with a mass of about 10^{18} grams, a diameter of about 10 kilometers, and hitting the earth at a speed greater than 50 kilometers per second. This paper provoked an enjoyable and productive controversy that has continued ever since. After almost two centuries, catastrophism is once more acceptable in geology, but what kind or kinds of catastrophe actually occurred? Lyell's uniformitarianism and Darwin's gradualism, though sound enough in their context, are no longer adequate to explain all the features we see at the ends of geological periods. In the meantime, we can all have fun vigorously contradicting each other, suggesting others' ideas are nonsense, and gradually developing a new scenario about just what did happen at the end of the Cretaceous or at other cusp points in geological time.

At the moment we are generally in agreement (or at least most of us are) that some kind of cataclysm marked the end of the Cretaceous and formed the Cretaceous-Tertiary Boundary, but not everyone agrees that this is what produced the extinction of the dinosaurs. Most workers in the field seem to think that the cataclysm was the result of an impact with an extraterrestrial body, though some cling to the idea that extensive vulcanism might explain the observations. Some believe that the impactor was an asteroid, and some opt for a comet. Many believe that mass extinctions have occurred on a strictly periodic timetable during the Phanerozoic, while others, including myself, believe that mass extinctions occur *on average* every 30 million years or so, with no measure of a regular period. Finally, there is no agreement about why an impactor (asteroid or comet) should actually hit the earth and cause the bang which wiped out so many creatures.

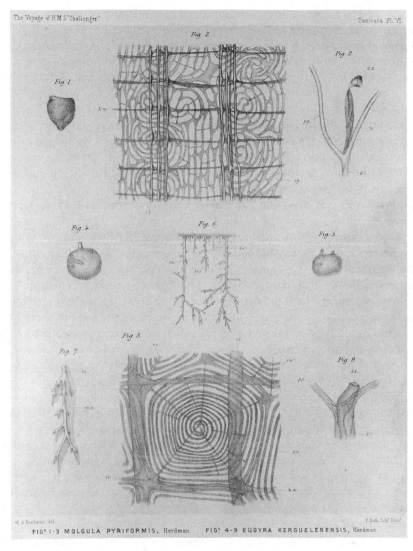

FIGS 1-3 MOLGULA PYRIFORMIS, Herdman   FIGS 4-9 EUGYRA KERGUELENENSIS, Herdman

Fig. 1.2. Protochordates, the tunicates *Molgula* and *Eugyra*. The entire animals are shown, about half life size, in figures 1, 4, and 5. The larger central illustrations represent the gills by which they filter their food from the water. Reprinted from the *Challenger Reports, Zoology,* vol. 6 (1882). (The *Challenger Reports* were the scientific results of the first round-the-world oceanographic survey by HMS *Challenger* between 1872 and 1876 under the scientific direction of Sir Wyville Thomson.)

Goethe, who was a fine scientist as well as a poet and playwright, wrote in his autobiography: "A false hypothesis is better than none at all. The fact that it is false does not matter so much. If, however, it takes root, if it is generally assumed, if it becomes a kind of credo admitting no doubt or scrutiny—that is the real evil, one which has endured through the centuries."

A false hypothesis is better than none at all. In one of his letters Darwin went further: "False facts are injurious to the progress of science, for they often endure long; but false views, if supported by some evidence, do little harm, for everyone takes a salutary pleasure in proving their falseness." How right Darwin was, in this as in so many things! Falsification of evidence is rightly regarded as one of the most shameful acts any scientist can perform and has been the subject of vigorous investigations in recent years.

Recent examples of falsification of evidence may be seen in the sad events surrounding David Baltimore's resignation, in the activities of N. Bonaventure in Paris and in the activities of V. G. Gupta in India. Baltimore, a Nobel Prize winner, seems to have succumbed to the temptation (a temptation to which too many senior scientists succumb) of putting his name as coauthor on a paper to which he had contributed little. He had put too much trust in the unsupervised work of a research associate who apparently had falsified data. Bonaventure has propagated results purporting to support homeopathic medicine, using impossible dilutions of materia medica and claiming effects from solutions that could only contain less than a thousandth part of a molecule per liter. The resulting scandal has been well aired in the pages of *Nature*. In India, Gupta, a senior government geologist, seems to have totally falsified the stratigraphy of the Himalayas by reporting fossils from there that he actually purchased in Morocco or elsewhere and never collected in the field. His many coauthors seem to have been deluded by him into accepting fossils he sent to them for independent study. Many of them have since recanted and dissociated their views from his.

Baltimore's offense was relatively minor (though that of his associate was not), a mere error of judgment and perhaps hubris, though one that has had serious consequences. Bonaventure's seems to have been largely self-deception, while Gupta's may be seen by some as extremely serious. How many more such falsifications of facts have occurred without detection? Perhaps more than we think, especially if the perpetrator made a good guess and merely fabricated data to support a hypothesis that later obtained independent genuine support and a measure of credence. One of the most famous examples is that of the Abbé Gregor Mendel. E. B. Ford and A. R. Fisher have shown that Mendel's assistants, who were chiefly novices of the monastery, falsified his data, apparently because they believed that they did not fit well enough Mendel's hypothesis, which later became the basis of genetics. They simply wanted to please the older man. False facts are indeed injurious to the progress of science. But what about false views, false hypotheses? In my opinion these can often advance the progress of science, provided they stir up controversy and cause everyone to take a salutary pleasure in proving their falseness. There is nothing like a good controversy to stimulate the search for new ideas, new supporting facts, new evidence. But beware of entering such a controversy unless you cannot merely accept but actually enjoy the give and take of vituperation!

This book is about the search for evidence to try to resolve the disputes about the Cretaceous-Tertiary Boundary—at least in my eyes. As William Blake wrote two centuries ago:

> I must Create a System, or be enslaved by another Man's;
> I will not reason or compare: my business is to Create.

I cannot create poems or paintings like his, but scientific creativity is the other side of the coin of artistic creativity, and I can create theories—and test them—and I plan to develop in this book a comprehensive scenario that, although it is expressed in words, is nevertheless firmly based in mathematical analysis. And thus stir up another storm of controversy.

Put briefly, I suggest that the Cretaceous-Tertiary Boundary event was a complex series of happenings, beginning with a nearby star turning supernova, in a gigantic cosmic bang. The first effect of this bang on earth was the arrival of the visible and near-visible radiation from the supernova, which produced ten or twenty times the heat of the sun at midday in the tropics, thus igniting worldwide forest fires. The blast from this explosion also perturbed the cloud of comets that surrounds the solar system, and some few centuries later one or more of these comets hit the earth, producing the well-known "nuclear winter" scenario and, more importantly, causing a tremendous acidification of the oceans.

I shall present evidence for each step of this idea—hard evidence for a nearby supernova, geochemical evidence for a comet (rather than an asteroid)—and show how this scenario provides a reasonable explanation of just which groups of animals and plants were wiped out in the Cretaceous-Tertiary Boundary event and which survived.

Now, on to the Badlands of Alberta, where dinosaurs once roamed the range—until diamonds fell from the sky!

2

The Badlands of Alberta

The valleys, which were the only practicable roads, for miles
and miles were not so much valleys as chasms or gorges,
sometimes two hundred yards across, but sometimes only
twenty, full of twists and turns, one thousand or four thousand feet deep, barren of cover, and flanked each side by pitiless granite, basalt and porphyry, not in polished slopes, but
serrated and split and piled up in thousands of jagged heaps of
fragments as hard as metal and nearly as sharp.
—Lawrence of Arabia, *The Seven Pillars of Wisdom*

Few people, indeed few Canadians,
know that Canada has a desert, and quite a large one at that. By all definitions of a desert—rainfall less than 100 millimeters (4 inches) per year
on average, vegetation, aridity of the soil—the Badlands of Alberta are
desert. The carrying capacity for cattle in the Sonora desert of southern
Arizona is about one head of cattle for every 50 acres, which is the maximum permitted on public lands by the U.S. Department of Agriculture in
that desert. In the Badlands of Alberta the corresponding figure is one
head of cattle for every 25 hectares (about 60 acres). On that basis a
10,000-acre ranch in the Sonora desert could support about 200 head,
while a similar ranch in the Alberta Badlands could only support about
160 cattle. The Badlands are more extreme desert than even the Sonora.
In uniform metric terms these figures become 48 cattle per square kilometer in the Sonora desert and 40 in the Badlands of Alberta.

The great cattle ranges of the Sudan, perhaps the paradigm of the African desert, can support even more cattle than the Sonora or than the Badlands of Alberta. Before the present civil war started there, the stocking
rate for the country as whole was about 25 hectares per head, or about
40 cattle per square kilometer, and this without evident overgrazing. The

stocking rate was not uniform, of course, throughout the country. The northern areas may see rain only once in 50 years; there, there are no cattle. The extreme south of the country is rain forest, now sadly much denuded, but there too cattle do not thrive, because of rinderpest and other diseases carried by the tsetse fly. It was north of the tsetse zone, on the short-grass savannah, and farther south, on the long-grass savannah, that the cattle people lived until their lives were disrupted by insurrection.

The Baqqara people, living along the southern fringes of the Sahara, within the short-grass savannah, but still technically in a desert area, since the rainfall is no more than an average of 100 millimeters a year, had stocking rates of up to 80 cattle for every square kilometer of land, twice that of the Badlands of Alberta. Indeed, in areas where they were free to undertake seasonal migrations, following the vegetation patterns, they could stock at more than twice this rate. The tribal wars with their southern neighbors, the Dinka of the semidesert tall-grass savannah, were never about grazing but more usually about access to water holes in the long dry season.

All three—the short-grass savannah of the southern Sahara, the cattle ranges of the Sonora desert in southern Arizona, and the Alberta Badlands—are desert, with the Badlands the least able of the three to support cattle.

The big difference in climate makes for large differences in vegetation, of course, but all three are basically short-grass prairie or savannah, much modified by overgrazing (especially in the Sonora) and punctuated with thorns—*Acacia* and *Zyziphus* (wait-a-bit thorn) in the Sudan, mesquite and cacti in the Sonora, and cacti in Alberta. Yes, cacti in Canada! Not the giant saguaros of Arizona, but ground-hugging cacti of several species that are able to withstand the winter cold, $-20°$ to $-30°C$ at times. And unlike Arizona, and most deserts of the world, the low temperatures ensure that surface water does not just evaporate.

Let me compare these three deserts—the Darfur desert of the Sudan (misleadingly called the Libyan desert, even that part of it which lies in the Sudan), the Sonora desert, and the desert of Alberta—by recalling my own entry into each one of them.

The Darfur Desert of the Sudan

The journey into the desert from Khartoum, itself in semidesert country (it receives all of 150 millimeters—6 inches—of rain a year) at the junction of the Blue and the White Nile rivers, starts north through countryside much like that described by Lawrence of Arabia—Jebel Surgham and the barren Kerreri Hills of "pitiless granite, basalt and porphyry." This is the battlefield where the British forces under Kitchener in 1898 defeated the Khalifa and revenged themselves for the murder of General Gordon some thirteen years before, as so well described by the young newspaper correspondent and lieutenant in the 21st Lancers, Winston Churchill.

Then the trail, for it is no more than a trail, turns west and enters the Qoz, the sandy desert, where I found it easier to travel on camelback than by truck, before reaching the hardpan plains of Dar Kababish, the short-grass savannah. Striking southwest across the plain, the trail leads once more to the northern edge of the Qoz, the grazing grounds of the nomadic tribes, the Baqqara, the once warlike Rizeiqat and the now decimated Ta'aisha. The landscape is dotted with *Acacia* shrubs, chiefly the flat-topped thorny *A. tortilis*, with occasional red-barked *talh* trees (*A. seyal*, the chief food of the giraffe in this area) and spinneys of the wait-a-bit thorn *Zyziphus*, with the recurved thorns that give it its common name. Near settlements the *Acacia tortilis* is replaced by *A. nubica,* the stinking thorn, which even goats will not eat. In the brief rainy season—no more than three weeks—the whole countryside is impassable. Much farther south, in the country of the Nilotic Shilluk, Nuer, and Dinka, one of the toughest regions in the world, the rainy season is longer, the villages are isolated, and the countryside—the *toich*—is so flat that the water does not drain away from the black cotton soil for months. It was here in the Sudd that great herds of longhorn cattle survived on the *toich* for centuries until the recent drought (a result of el Niño) and the revolution scattered them into Ethiopia and Uganda, back to desert conditions.

In the land of the Baqqara (Figure 2.1), Dar Baqqara as it is called by the people, the desert nomads, by trekking north onto the hardpan in the rainy season and south onto the savannah in the dry, could until recently stock the land at a rate of about one head of cattle per four hectares, or 250 per square kilometer, roughly one for every ten acres, five times the stocking rate of the Sonora but a quarter that of the *toich*.

The Sonora Desert of Arizona

Driving southwest from Tucson into the Sonora through the Altar valley there is hardly any flat land (Figure 2.2). The valley bottom is rolling grassland, with the worst overgrazed areas dotted with creosote bushes and mesquite, and the tall saguaros grow slowly on the hill slopes. Only in the game reserve, operated by the federal government, at the southern end of the Altar valley is the grassland gradually returning toward its pristine condition, that of short-grass prairie. High ranges of hills and mountains rise on either side: Kitt's Peak, with its astronomical observatory, and Baboquivaro mountain, sacred to the Tohono O'odham Indians, who believe it is the center of the earth and home of the creator I'itoi. The surface of the valley is rocky and broken, and once off the road a surefooted horse is the best transport. For most of the year the rivers and streams are dry sand "washes," only bearing water for a few weeks in the monsoon season. Once there is water, though, the brilliant colors of desert flowers are enough to make even a geologist look up from his rocks and see life at its freshest.

The Badlands of Alberta

Driving east from Calgary towards the Badlands (Figure 2.3), the last outlying hills of the Rockies are soon left behind and the landscape becomes a flat plain, almost a mile above sea level. At first this is semidesert, able to provide a wheat crop without irrigation. Soon the landscape changes to irrigated fields of canola, a hybrid oil-seed crop developed in this region, and the more recently developed hybrid flax, also grown for its oil more than for its fiber. In the spring the patchwork of yellow canola fields and blue flax fields makes a striking pattern. Surface water is rare, but what streams there are carry water all year round. Gradually, even the irrigated crops disappear, and the landscape is flat cattle country, with little to be seen from the road except grass and an occasional mule deer. Even when it rains, the few wildflowers are inconspicuous.

Centuries ago this was the hunting ground of the Blackfoot, one of the first nations of Canada, and of the buffalo, which were their chief sustenance. Here they hunted the buffalo; here they fought with their enemies from the north, the Cree, and from the west, the Kootenai, the Sarcee, and the Dakota Sioux. Great herds of buffalo (bison really, but who cares?) roamed across these plains, followed by the nomadic tribes of the Blackfoot nation.

We drive on across the waving plains of grass, then suddenly, around a bend, a deep gash opens in the ground, like a miniature Grand Canyon, a few hundred feet deep, and the Badlands finally justify their name. This is an inverted landscape, a flat plain with canyons instead of mountains, canyons with the soft siltstone and mudstone carved into fantasies of sculpture by running water, rain, snow, and wind, shapes locally called "hoodoos." These were the cliffs where the Blackfoot hunters drove the herds of buffalo over the edge. One such site has been reconstructed at a location known as Head-Smashed-In Buffalo Jump, where there is a magnificent museum of Piegan life; the Piegan are one sept of the Blackfoot nation.

Fig. 2.1. Dar Baqqara, the country of the nomadic Baqqara peoples, lies along the southern fringe of the Darfur desert of the Sudan.

Fig. 2.2. The Sonora desert lies along both sides of the border between Arizona and the state of Sonora in Mexico.

Fig. 2.3. The Badlands of Alberta lie in the eastern part of southern Alberta.

Fig. 2.4. The Red Deer River valley as it was in 1884. This photograph was taken by J. B. Tyrrell in what is now the Dinosaur Provincial Park, a hundred kilometers south of the major study site for the research in this book. Printed from the original glass negative, by permission of the Thomas Fisher Rare Book Library, University of Toronto.

The Red Deer River Valley

The former mining town of Drumheller can be considered the center of the Badlands, located on the Red Deer River, near where several creeks augment the flow—Rosebud River, Ghostpine Creek, and Kneehills Creek from the west, and Michichi Creek from the east. The town is situated several hundred feet below the plain, sheltered from the constant prairie wind, and has taken a new lease on life since the Royal Tyrrell Museum of Palaeontology was opened there in the 1980s.

South from Drumheller, past Deadhorse Lake, Dinosaur Provincial Park encloses more than 50 percent of the known species of Cretaceous dinosaurs that have been found so far. The Red Deer River and its tributaries have cut down through hundreds of feet of Cretaceous strata, holding, like a chest of drawers, a rich cache of dinosaur remains (Figure 2.4). Unlike most rocks, which were laid down under the sea, the strata here were deposited on land and have suffered little disturbance—no folding or heating by mountain-building forces, no compression, hardly even any tilting. Certainly they have been elevated in the 70 million or so years since they were formed, but the elevation has been even and gradual. The strata tilt no more than one degree from the horizontal.

Lacking any history of compression, heating, or folding and never having been under water for long periods, the rocks in this area have never been consolidated. In fact they hardly deserve to be called rocks at all. I have seen dried silt from the Nile River harder than some of the Cretaceous rocks of the Alberta Badlands. It is this very softness of the rocks that makes it so easy to find fossils, since the mudstones and sandstones weather away, exposing the harder fossils.

Because of the tilt of the strata, slight though it is, the farther south one goes along the Red Deer Valley, the older are the rocks. At Drumheller the youngest rocks, at the top of the canyon, are about 68 million years old. In Dinosaur Provincial Park the youngest rocks are over 70 million years. Correspondingly, moving north from Drumheller we encounter progressively younger rocks until, near the village of Scollard, the surface of the plain consists of Tertiary Paleocene strata about 64 million years old. At this point the canyon of the Red Deer cuts right through the Cretaceous-Tertiary Boundary, the moment when the great extinction occurred that terminated the Mesozoic era.

In the wall of the Red Deer River canyon in this neighborhood, the Cretaceous-Tertiary Boundary is exposed as a thin layer of rock, like a thin layer of butter between two slices of bread, to use the analogy of William Smith, an eighteenth-century engineer to whom we owe the basis of stratigraphical science. Below the Cretaceous-Tertiary Boundary the rocks are replete with microfossils characteristic of the Cretaceous period: pollen grains, spores of ferns, and microorganisms. The Cretaceous-Tertiary Boundary layer itself is devoid of fossils of any kind. Above the Cretaceous-Tertiary Boundary layer, the rocks at first have sparse microfossils of species characteristic of the Paleocene period, the lowest period of the Tertiary era; then numbers and diversity, especially of fern spores, increase. The rock layer immediately above the boundary is a layer of coal half a meter thick, indicating abundant vegetation but with no signs of more than a very few species of plants. No evidence of dinosaurs is found above the Cretaceous-Tertiary Boundary nor, in fact, of any preexisting land animals larger than about 25 kilograms, the size of a small dog.

The approach to the canyon, after obtaining permission from the farmer who owns the land on the plain, is by four-wheel-drive vehicle to near the edge. Then comes a short walk across waving grassland and a slither down the wall of the canyon to the requisite level. It's easy to overshoot, but ledges of slightly harder rock help to prevent too long a slide (as I know only too well). The challenge lies in carrying geological and survey equipment down a steepish cliff face, one that is too soft for any kind of climbing holds. Did I say science is fun? It is, you know. Working on the face on a fine day can be most enjoyable and invigorating, sheltered from the constant wind of the prairie, warm enough for constant vigorous work but not too hot in this cold climate. The river running a few hundred feet below, the gargoyle shapes produced by erosion, the physical workout provided by the site all contribute to a feeling of well-being. A bull moose feeding in a backwater of the river far below, elk and a small herd of mule deer in the thickets of the valley bottom, hawks soaring on the updrafts, even an occasional gyrfalcon contribute to a feeling of fellowship with my coworkers in the field and of oneness with the environment. This is no national park, so work is undisturbed by the stream of curious tourists who constantly interrupt paleontologists in their labors in the Dinosaur Provincial Park farther south. Sometimes I wish I had brought a fishing rod with me, but when I look at the cliff face—down *and* up for several hundred feet—I realize that my fieldwork is more enjoyable than spending an hour fishing. Leave fishing to the bears. I'll take the Cretaceous-Tertiary Boundary any day!

It was the butter of the Cretaceous-Tertiary Boundary between the slices of bread of the Cretaceous rocks below and the Tertiary rocks above that brought me to this remote location in the Badlands of Alberta in the company of Dr. Denis Braman from the Royal Tyrrell Museum of Palaeontology in Drumheller. At a casual glance the cliff face seemed almost uniform in color and texture, but on close examination and survey, subtle differences became clear. We surveyed the face for 5 meters above and below the Cretaceous-Tertiary Boundary. Below, the face consisted chiefly of layers of mudstone and siltstone, hardly firm enough to be called rock, but with abundant microfossils. The Cretaceous-Tertiary Boundary layer itself was about 8 to 10 millimeters thick, consisting of a gray claystone. Immediately above was a stratum of soft—very soft—coal, about half a meter thick, then more layers of siltstone and mudstone, this time with relatively few microfossils, then a layer of bentonite or volcanic ash from a long-dead volcano, more siltstone, a thin seam of soft coal, then more mudstone and siltstone layers.

The surface was badly weathered, and the strata only became evident when we cleared away the surface material with a trowel. In fact, the rock was so soft that all we needed in order to collect samples at various levels was not a geologist's hammer but a simple bricklayer's trowel. These rocks, now exposed in desert country on the face of a cliff in the canyon of the Red Deer River, had been laid down in lowland marshlands around the fringes of the continent, land that was no doubt dry in summer, but

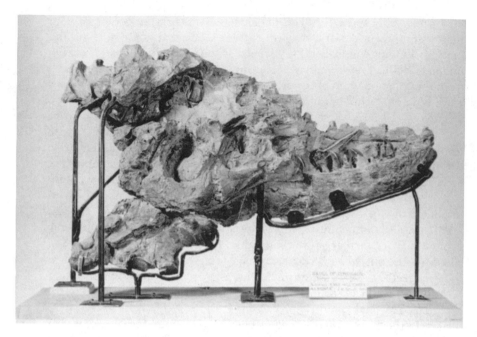

Fig. 2.5. Albertasaurus skull. The original specimen, which is about a meter long, was discovered by J. B. Tyrrell in his 1884 expedition. Printed from the original glass negative, by permission of the Thomas Fisher Rare Book Library, University of Toronto.

frequently flooded by freshets bringing silt and mud down from the mountains to the west. The delta of the Nile River in Egypt might perhaps be comparable today, though that is surrounded by desert. A better comparison might be the whole country of Bangladesh, low-lying land, much of it marshland until drained by human endeavor, a land subjected to heavy monsoon rains and to droughts in the dry season. Here, in the delta of the Ganges and Brahmaputra rivers, remain numerous flat, shifting fertile islands, built up from the silt carried down toward the sea by these great rivers. In comparable conditions in Alberta in the late Cretaceous roamed the herds of herbivorous dinosaurs preyed on by such creatures as *Albertasaurus* (Figure 2.5), discovered by J. B. Tyrrell after whom the museum is named, during his geological survey of Alberta in 1884.

Then suddenly the dinosaurs were no more. Where once they had kept the vegetation to a close-cropped sward, not of grass but of low-growing plants, especially ferns and cycads, there was a complete change. What had happened? Among the survivors of the Cretaceous-Tertiary Boundary event, fern spores seem to have been the most successful in this locality, and immediately above the boundary clay we find a coal layer formed from the remains of a forest of tree ferns. Once there were no dinosaurs left to crop the vegetation close, the surviving fern spores could develop into much larger, taller plants, not plants with deep root systems like present-day trees, but surface-hugging fine roots that did not disturb the underlying soil to any great depth, doing little more than drawing forth water and nutrients.

This relatively undisturbed Cretaceous-Tertiary Boundary layer, laid down on land, not under the sea, offered far better prospects for detailed chemical studies than the better-known Cretaceous-Tertiary Boundary rocks from Italy, Spain, France, New Zealand, Texas, Turkmenia, and many other localities, all of which were deposited under the sea, consolidated by heat and pressure, and twisted and deformed by mountain-building processes. Here in the Badlands of Alberta I felt I could find and sample the rocks with which I could best test the ideas I was developing about the events that occurred at the Cretaceous-Tertiary Boundary.

What the Cretaceous-Tertiary Boundary Layer Reveals

The Cretaceous-Tertiary Boundary marks the end of the Mesozoic era as well as of the Cretaceous period. Below this layer the rocks are full of Mesozoic fossils, dinosaurs and ammonites, cycads and chalk-forming organisms. Indeed, the very name Cretaceous (from the Latin word *creta*, for chalk) referred in the first place to the White Cliffs of Dover, that chalk formation par excellence, whose formation came to an abrupt halt at the Cretaceous-Tertiary Boundary. The Cretaceous-Tertiary Boundary clay itself is peculiar, not like any common clay in composition, but possessing the relatively uncommon, but by no means rare, mineral smectite instead of the illite of common clays. This clay is totally lacking in fossils of any kind.

Above the Cretaceous-Tertiary Boundary, the rocks at first contain very few species of fossils, though sometimes those few are very abundant, as in the coal of the Red Deer locality. Then the number of species increases, not the number of the old forms present below the boundary, but of totally new species and new groups. No dinosaurs, pterosaurs, ichthyosaurs, or mosasaurs; no ammonites or chalk-forming organisms. The fossils of planktonic animals and plants were slow to appear in rocks laid down under the sea, so that over a period of a million years or more we find a slow and gradual increase of numbers of fossils as well as diversity as we progress up the rock column. This is quite different from the condition we find in the Red Deer exposure, where abundant ferns immediately led to the formation of coal above the Cretaceous-Tertiary Boundary, but even here the diversity remained low for a million years.

The events occurring at the Cretaceous-Tertiary Boundary form the subject of this book. What caused the sudden disappearance of so many forms of life, if indeed the disappearance was sudden? Why did some species survive while others were wiped out? What of other mass extinctions at other epochs—were they also caused by the same kind of mechanism? Were mass extinctions periodic? Is there anything but a quantitative difference between the extinctions that terminated the Paleozoic and the Mesozoic eras and those that terminated, in a less spectacular fashion, the lesser geological periods, or is there a qualitative difference too? In other words, were all mass extinctions much the same except in magnitude?

The resolution of these questions is an ongoing process, with many false starts and much frantically wrong theorizing. Nevertheless, theories—however fragmentary, incomplete, misleading, or just plain wrong—are necessary, if only to suggest just what further observations might be worth making. Making an observation of any kind in this field is often physically demanding, time consuming, and laborious, often only worth doing if there is some kind of theory to test. For instance, since everybody "knows" that diamonds only occur in association with a certain type of volcanic rock, why would we go to all the trouble to look for diamonds in the Cretaceous-Tertiary Boundary rocks, especially if the diamonds we expect to find are so minute—a billion billion or more to a carat—as to be of no possible commercial value? We do it because comets contain just such diamonds so that the presence of these extraterrestrial diamonds in these rocks might provide evidence that a comet hit the earth at this time.

And, in fact, we have found these minute diamonds in the Cretaceous-Tertiary Boundary rocks of the Badlands of Alberta.

Death and Survival at the Cretaceous-Tertiary Boundary Event

And all flesh died that moved upon the earth, both of fowl, and of cattle, and of beast, and of every creeping thing that creepeth upon the earth. . . . All in whose nostrils was the breath of life, of all that was in the dry land, died. And every living substance was destroyed which was upon the face of the ground. . . . and they were destroyed from the earth.
—Genesis 7: 21–23

Dinosaurs have captured the public imagination to such an extent that even people who ought to know better tend to regard the Mesozoic era as the Age of Reptiles or the Age of Dinosaurs. In fact, dinosaurs seem to have been quite scarce, both in numbers of species and numbers of individuals. There is no evidence that there were ever more than 25 species of dinosaurs alive at any one time, though imperfections in the fossil record suggest that this is a somewhat low estimate; some authorities have suggested that at their peak there may have been as many as a hundred species.

Even that most renowned of all dinosaurs, *Tyrannosaurus rex*, seems to have been a rarity. Until the recent discovery of a new nearly complete specimen, only three reasonably complete fossil skeletons were known, together with a number of scattered bones. The king of the Cretaceous jungle seems to have been much rarer than the king of beasts today.

The Mesozoic: The Age of Mammals

The herds of herbivorous dinosaurs cannot begin to compare either in numbers of individuals or variety of species with the antelopes of the present day, let alone with the vast array of herbivorous rabbits and marmots, pica and squirrels, deer and elephants. *Tyrannosaurus rex* may have been huge, but also rare, and his relatives amongst the carnivores were few. The Mesozoic had nothing to compare with the great cats and the canids of the present day, either in numbers of individuals or in numbers of species, lions and jaguars, wolves and hyenas. The number of species, both of herbivorous and of carnivorous mammals alive today, greatly exceeds the total of herbivorous and carnivorous dinosaurs alive, not just at any one time but for the whole of the nearly 200 million years of the Mesozoic.

Though dinosaurs existed only during the Mesozoic, we cannot call this the Age of Dinosaurs; other creatures were far more abundant. Indeed, we know more species of mammals from the Mesozoic than we do reptiles, while today there are more species of reptiles—including snakes, lizards, crocodiles, and turtles—than there are of mammals. It might, in fact, be more correct to call the Mesozoic the Age of Mammals and the present day the Age of Reptiles!

Arrays of footprints made in the mud of lakeshores and now preserved in rock as fossils suggest that some species of herbivorous dinosaurs moved as family groups, rarely in anything like the herds of gnu that now cross the plains of Africa in their millions. I have seen a steady stream of migrating antelopes crossing the White Nile River in the Equatoria province of the Sudan, continuing for days on end (and this is not even one of their major migration routes). Such migrations leave no distinct trails of footprints in mud, but instead leave deeply eroded lanes where individual footprints are indistinguishable. Nothing like this is known from the Mesozoic.

Large bones, and especially teeth, survive the process of fossilization far better than do small fragile bones, being less easily dissolved by acids in the soil or by bacterial action. Yet in the Cretaceous period teeth of tiny, fragile mammals are more abundant than are the teeth of dinosaurs.

The dinosaurs were spectacular, but they were by no means the most abundant of the land vertebrates, either in numbers, in number of species, or even probably in biomass. We find evidence in fossil footprints of hadrosaurs moving in family groups and even nesting communally, but where is the evidence of herds of herbivorous dinosaurs equivalent to the bison herds that once covered North America or the herds of antelopes in Africa? Even the herds of African elephants (before human slaughter decimated them) seem to be larger in numbers than anything we can infer from the fossil record about herds of dinosaurs.

We must remember, however, that terrestrial fossils are rare compared to marine fossils. Fossilization is only really possible in areas where sediments are accreting, not where the rocks are eroding. The coast of northern California around Big Sur, the cliffs of Oregon, the icy inlets of Alaska, and the fjords of Norway, where the rocks are eroding under the impact of tides and glaciers, are no place to expect fossils to be formed. The accumulating sediments of Chesapeake Bay or of Lake Erie, in contrast, consist in great degree of planktonic fossils. We can determine the moment of when "white man's" agriculture began by searching for the "*Ambrosia* horizon" in the sediments of Lake Erie, that level in the sediments where ragweed (*Ambrosia*) pollen first becomes abundant. There is no place in the eroding peaks of the Rocky Mountains where fossilization could take place, and the great plains of the prairies too are eroding—witness the canyon of the Red Deer River, where we have been working on the Cretaceous-Tertiary Boundary. The peat bogs of Ireland consist entirely of subfossil plants and a few animal remains, and if they were not currently being cut for fuel (Ireland even has one power station that uses peat), they would be future terrestrial fossil beds. Indeed, these peat beds have been the origin of most of the known giant elk remains, subfossil specimens of *Megaloceros* from the Pleistocene.

The Landscape Before the Cretaceous-Tertiary Boundary Event

The fossil dinosaur beds of North America represent the marshy lowlands around the fringe of a continent. The mountainous interior, especially the western margins, of that continent was eroding and offered no possibility for the preservation of fossils. It is quite possible, though unlikely, that mountain-dwelling goatlike dinosaurs lived there and left no trace for us to find in the bone beds. The reason this is unlikely is because some of the bone beds we know from this period represent the remains of creatures that fell into rivers and were swept down to be preserved in sandbanks in the bends of these rivers. Such bone beds include fish remains as well as mammalian teeth and dinosaur remains, though the more robust bones predominate, often rolled and ground down by the action of running water and gravel; most mammalian bones would not survive such treatment.

By contrast, the Cretaceous fossil beds of the Dinosaur Provincial Park in Alberta, Canada, contain remains that were fossilized in situ, not swept downstream from above. Some bones are articulated, giving a more-or-less complete skeleton, but most are scattered single bones and especially teeth. Careful sieving of the sediments yields more mammalian teeth; dinosaur teeth are usually visible to the naked eye without any need for sieving. Here were the seasonal marshes, flooded in the spring by freshets from the mountains, dry for part of the year, accumulating sediment and hence preserving fossils. Rank vegetation, much of it ferns and cycads, was browsed down by herbivores, both the large dinosaurs and smaller creatures, including mammals and especially insects. Scattered trees and occasional copses dotted the landscape, but there is no sign of general dense forest (at least this far north), perhaps because the abundance of browsers prevented seedlings from developing into trees. Fossil pollen grains, seeds, and spores by the billion make up a large percentage of the sediments in these places.

Though this was far from the dominant landscape before the Cretaceous-Tertiary Boundary, it is the one we know most about, for of all types of terrestrial landscape this was the one that was most favorable for fossilization of terrestrial flora and fauna. By contrast, we know far more about the underseascape. We can know nothing about conditions of life on an erosional coast, for no fossils can form there as they can on a depositional coast. But offshore from an erosional coast the material eroded from the cliffs and beaches is deposited and may contain fossils, both of plankton (the floating creatures) and nekton (the fast-swimming animals) and of the benthic creatures like oysters that live attached to the seabed. Once more the dominant fossils, both in biomass and in numbers, are the tiny creatures, the plankton. Indeed, some rocks formed at this time consist entirely of fossil planktonic organisms, including the chalk from which the Cretaceous gets its name. The coccolithophorids (Figure 3.1), planktonic algae whose skeletons, together with those of the foraminiferans that ate them, largely formed the White Cliffs of Dover, were the most abundant planktonic organisms of this period.

What Groups Survived and What Did Not

In considering which taxa were exterminated at the Cretaceous-Tertiary Boundary, we are faced with five groups to consider: bacteria; marine plankton (both animal and plant); large aquatic animals, including fish; terrestrial plants; and terrestrial animals.

Fig. 3.1. Prisius dimorphosus, a coccolithophorid extracted from a core taken into the Maud Rise in the Weddel Sea. The individual coccoliths are the more-or-less circular objects that cover the surface of the plant. This species first appeared about 500,000 years after the Cretaceous-Tertiary Boundary. The white scale bar is 1 micrometer long. Reproduced by permission of James J. Pospichal, University of Florida.

Bacteria

Fossil bacteria are surprisingly common in the rocks below the Cretaceous-Tertiary Boundary in Alberta, rocks laid down under terrestrial conditions. They are rarer in marine rocks. None at all has been found in the boundary clay itself, but that does not mean that these organisms were exterminated, but rather that the claystone was formed by deposition from the atmosphere in a short period of time. Indeed, it is no more than a surmise that large numbers of bacteria were killed, and there is no evidence at all that any major taxa (above orders) were destroyed by the Cretaceous-Tertiary Boundary event. We do know that all major taxa of bacteria at the present day far antedate the Cretaceous-Tertiary

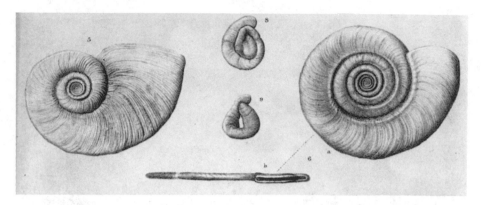

Fig. 3.2. Cornuspira species, a modern foraminiferan. Reprinted from the *Challenger Reports, Zoology*, vol. 9 (1884).

Boundary event, and there is no sign in the fossil record that any major groups of bacteria existed before that event that we do not find today. In other words, though some species of bacteria may have been wiped out, no major taxa were extinguished and no new major taxa evolved.

Marine Plankton

The story is far different when we come to consider the marine plankton. The two most important groups of algae and of protozoa, respectively, were the coccolithophorids and the foraminifera (Figure 3.2), whose skeletons form the greater part of the Cretaceous chalk. Eighty-five percent of the coccolithophorid species extant in the topmost Cretaceous did not survive the Cretaceous-Tertiary Boundary event, while 92 percent of foraminiferan species were wiped out. Both groups have since diversified again. At the present day coccolithophorids are once again forming chalk in tropical waters, and some formed new chalk deposits as early as the Eocene epoch; most foraminifera today are bottom-dwelling organisms, not planktonic at all.

Both these groups of planktonic creatures have calcareous skeletons that are readily fossilized. Indeed, in most areas of the continental shelf today any of these organisms that have not been eaten by something larger will settle to the bottom and will almost without exception be fossilized.

But many species of planktonic algae have no hard parts that could be fossilized, so we have no record of how they fared through the Cretaceous-Tertiary Boundary event. Did *Volvox* (a green alga with no hard parts) originate before or after this event? Is it an evolutionary result of the Tertiary, or did it survive the Cretaceous-Tertiary Boundary event? We can be certain that the Cyanobacteria (or Cyanophyta, depending on whether you are a microbiologist or a botanist; blue-green algae to you and me) are an ancient group, antedating the Cambrian explosion.

Two groups of algae gave a mere hiccup at the Cretaceous-Tertiary Boundary and then became far more abundant than they had ever been: diatoms (Figure 3.3), with their siliceous skeletons, today dominate the colder waters of the globe and share dominance of the temperate waters with the second group, the dinoflagellates (Figure 3.4), some of which are notorious as the red-tide organisms responsible for major fish kills and for paralytic shellfish poisoning. Both these groups of algae form resting spores to survive harsh conditions. Their spores are abundant in deposits formed both before and after the Cretaceous-Tertiary Boundary event but only come to dominance after this event.

Planktonic animals with a siliceous skeleton did not fare as well as these siliceous algae. A mere 15 percent of species of radiolarians (Figure 3.5) survived the Cretaceous-Tertiary Boundary event, perhaps because many of them depend on symbiotic algae for their food (Figure 3.6), but the few survivors rapidly diversified until today their skeletons form the bulk of deep-sea sediments, the so-called radiolarian ooze.

Large Aquatic Animals

Most larger planktonic herbivores have few hard parts, making it difficult to assess how well they survived. The protochordate salps (Figure 3.7), doliolids (Figure 3.8), and larvaceans (all tunicates, distant relatives of the vertebrates) certainly date from the Paleozoic, though fossils are practically nonexistent. Somewhat better represented in the fossil record are the planktonic crustaceans, thin (and hence less easily fossilized) though their carapaces may be compared with their bottom-dwelling cousins. The dominant group in the Cretaceous seems to have been the ostracods (Figure 3.9), which are relatively rare today, while copepods (Figure 3.10), which dominate the ocean fauna nowadays (and account for some 90 percent of the total animal biomass of the earth, being far more abundant than krill or insects), were rare before the Cretaceous-Tertiary Boundary event.

None of the major taxa we have been considering so far were exterminated in their entirety—decimated perhaps and so allowing other groups to take over after the Cretaceous-Tertiary Boundary event, but not totally wiped out. Individual species, genera, and whole families were eliminated within these groups, but not larger, more-encompassing taxons such as an

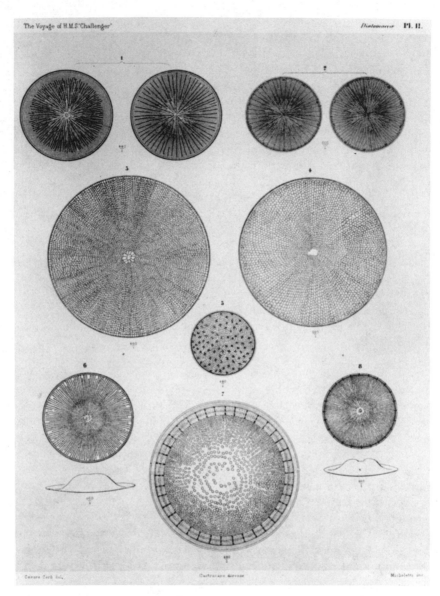

Fig. 3.3. Diatoms, single-celled plants with siliceous skeletons. Reprinted from the *Challenger Reports, Botany*, vol. 2 (1886).

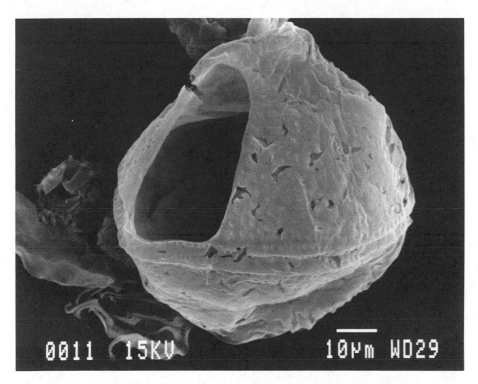

0011 15KV   10µm WD29

Fig. 3.4. Dinoflagellate cyst (dormant stage) of *Apteodinium spiridoides*, from the lower Miocene. The white scale bar is 10 micrometers long. Reproduced by permission of M. J. Head and the American Association of Stratigraphic Palynologists Foundation, from Head and Wrenn (1992), pl. 8, fig. 1.

order or class. It is when we turn to larger marine animals, especially the predators, that we first find whole orders extinguished.

The most prominent order that was totally wiped out in the sea was the ammonites, pelagic molluscs related to modern squid (Figure 3.11) but possessing a flat spiral shell of enormous complexity and often of large size—as big as a cartwheel sometimes; rare specimens are known that are 3 meters across (Figure 3.12). Inner compartments of the shell were filled with gas and served as a flotation device. If their modern relatives are anything to go by, they were among the most intelligent creatures in the sea. Not a single species of ammonite survived the cataclysm of the Cretaceous-Tertiary Boundary event. The belemnites, a carnivorous taxon even more closely related to the squid than the ammonites, being effectively squids with straight shells filled with gas as a flotation device, were also extinguished in the Cretaceous-Tertiary Boundary event (Fig-

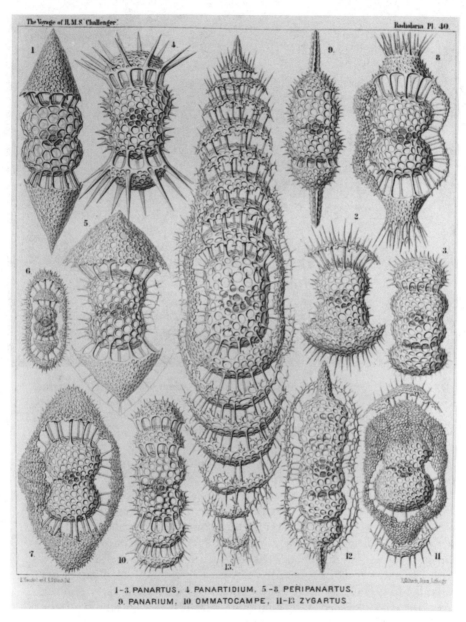

Fig. 3.5. Radiolaria, single-celled animals with siliceous skeletons. Reprinted from the *Challenger Reports, Zoology*, vol. 18 (1887).

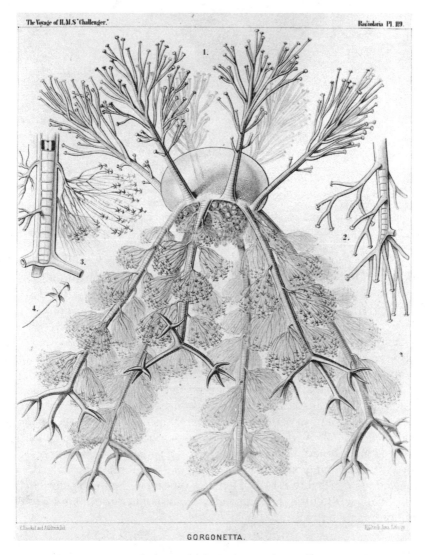

GORGONETTA.

Fig. 3.6. Gorgonetta, a radiolarian with symbiotic algae. This species has a spiny siliceous shell in the form of a bell. In sunlight the mass of symbiotic algae emerge through the opening of the bell and commence photosynthesis. Reprinted from the *Challenger Reports, Zoology*, vol. 18 (1887).

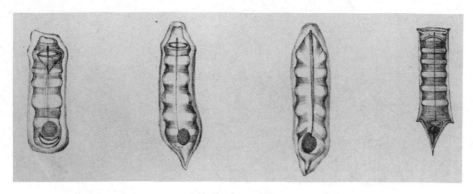

Fig. 3.7. Salpa zonaria, a soft-bodied creature, related to vertebrates and unknown from the fossil record. The biological evidence indicates that it evolved long before the Cretaceous-Tertiary Boundary. Reprinted from the *Challenger Reports, Zoology*, vol. 17 (1887).

ure 3.13). Related groups, known collectively as the cephalopods, were decimated but survived, the shell-less squid most successfully, but also the nautiloids (there are five species of this taxon alive today, see Figure 14.2), the bottom-dwelling cuttlefish and octopods, and, most strangely, the argonaut with its fragile shell. (The shell of a female *Argonauta*—the male has no shell—collected in the Mediterranean, is the most fragile specimen in my own private collection.)

Bottom-dwellers fared better than pelagic animals. Many families of clams and sea snails were exterminated, but no whole orders. No more than about 40 percent of these groups were wiped out. Brachiopods, looking rather like clams but quite unrelated, survived somehow; an earlier mass extinction had nearly eliminated this group, but the few species of survivors have hung on to the present day in small numbers. Indeed, one genus, *Lingula* (Figure 3.14), which first appears in the fossil record in the Cambrian, seems little different today from its ancestors found as 400-million-year-old fossils.

Terrestrial Plants

Cycads were prominent in the Mesozoic forests. Today only a single family remains as localized rarities of the tropics, whose large and somewhat poisonous nuts provided food for the people of Guam during the Second World War. The toxin is a slow-acting neurotoxin (an unusual amino acid) that can produce irreversible brain damage, known as Guam disease. It is interesting to speculate whether cycad nuts were eaten by

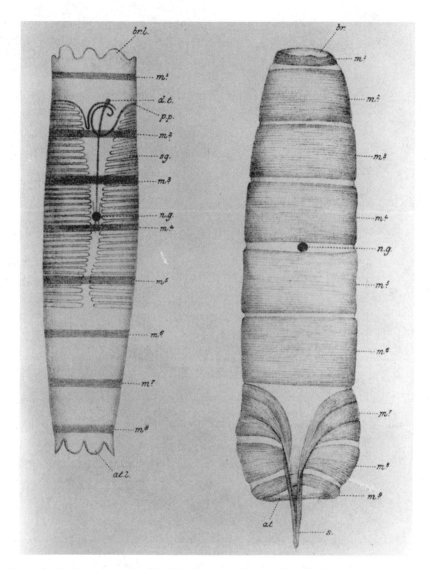

Fig. 3.8. *Doliolum* species, like *Salpa*, evolved long before the Cretaceous-Tertiary Boundary but are quite unknown from the fossil record. Reprinted from the *Challenger Reports, Zoology,* vol. 17 (1887).

W.Purkiss lith

1 a..d CYTHERE ERICEA, Sp.nov.   3 a..d CYTHERE NORMANI, G.S.Brady
2 a..d   .   IRPEX, Sp.nov.   4 a..f   .   DASYDERMA, Sp.nov.
          5 a..f CYTHERE SCABROCUNEATA, Sp.nov.

Fig. 3.9. Ostracods, tiny bivalved crustaceans, were the dominant crustacea before the Cretaceous-Tertiary Boundary but are now relatively uncommon compared with copepods. Reprinted from the *Challenger Reports, Zoology*, vol. 1 (1880).

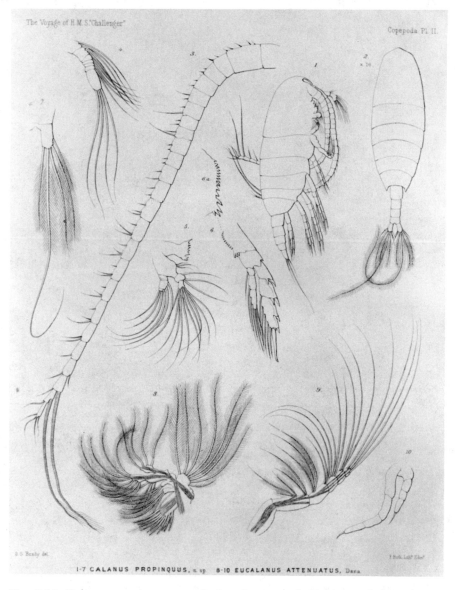

Fig. 3.10. Calanus propinquus, a planktonic copepod. Unknown before the Cretaceous-Tertiary Boundary, these are now the dominant animals in the sea. Reprinted from the *Challenger Reports, Zoology*, vol. 8 (1883).

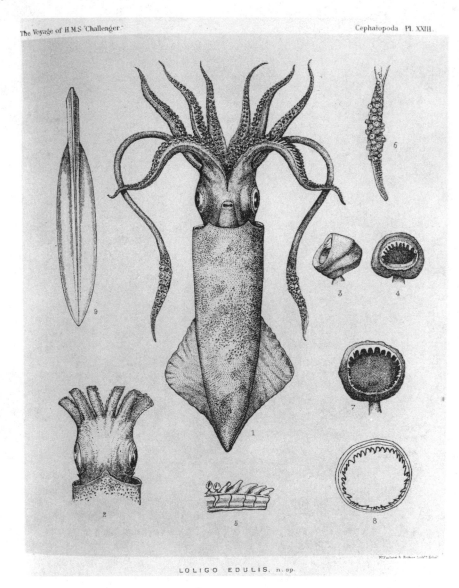

Fig. 3.11. *Loligo edulis*, the edible squid. Reprinted from the *Challenger Reports*, *Zoology*, vol. 16 (1886).

Fig. 3.12. Lioceras concavus, an ammonite. Reprinted from the *Palaeontographical Society Monographs*, vol. 41 (1888).

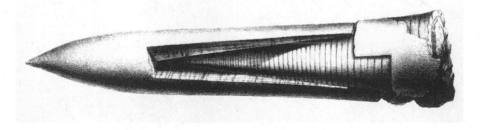

Fig. 3.13. Belemnites abbreviatus, a belemnite. Reprinted from the *Palaeontographical Society Monographs*, vol. 23 (1870).

dinosaurs during the Mesozoic and, if so, what effect the poison had on them.

Ferns survived far better than cycads, and the forests of the early Tertiary period were formed largely of tree ferns, mingled with conifers in northern latitudes. The few species of coniferous trees present in the Cre-

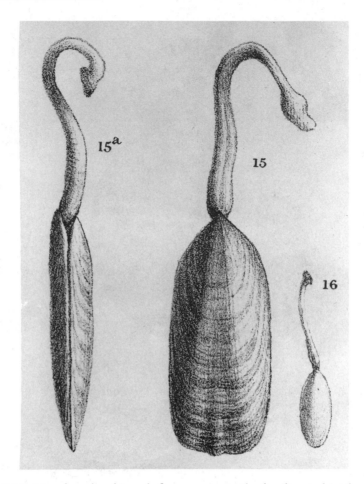

Fig. 3.14. Lingula, a brachiopod, first appears in the fossil record in the Cambrian, 400 million years ago, but is still living today. Reprinted from the *Challenger Reports, Zoology,* vol. 1 (1880).

taceous forests seem to have been exterminated, and the conifers, which today dominate the boreal forest, survived the Cretaceous-Tertiary Boundary event as small heathlike plants, as did the flowering plants that later gave rise to the great trees of the temperate forest and to the grasses of the prairies, steppes, and suburban lawns.

Extinctions of land plants were fewer in northern latitudes than in warmer southern ones, probably because northern plants have developed mechanisms for dormancy that enable them to overwinter. Modern cycads have no such dormancy mechanisms, but some fern spores may survive long periods. Such mechanisms would have enabled their possessors to survive the Cretaceous-Tertiary Boundary cataclysm. Thus, in the southern forests of North America (tropical rain forests in effect, even though outside the tropics), 75 percent of species of large-leaved plants were extinguished, while in Alberta 70 percent seem to have survived.

Terrestrial Animals

No land animal heavier than 25 kilograms is known to have survived. No dinosaur is known that could have turned the scales at less than 70 kilograms (about 155 pounds). Over the magic weight of 25 kilograms, the only animals that survived lived in water: crocodiles and turtles. But other aquatic reptiles perished like the dinosaurs: crocodilelike Mosasaurs, long-necked Plesiosaurs, short-necked Ichthyosaurs—three whole orders exterminated at the Cretaceous-Tertiary Boundary event. The amphibians—frogs and toads, salamanders and newts—fared better.

Of the airborne creatures, the pterodactyls were wiped out, but the birds survived. Bats had not yet evolved. The records of insects are sparse, and we can determine little about them. All orders of present-day insects, however, seem to have evolved before the Cretaceous-Tertiary Boundary event.

In total, six whole orders of reptiles—two orders of dinosaurs, three of aquatic reptiles, and one of aerial pterosaurs—were exterminated by the Cretaceous-Tertiary Boundary event, far more than of any other class of animals. Among the vertebrates, the reptiles were the great losers in the Cretaceous-Tertiary Boundary cataclysm, and in this sense it is perhaps fair to say that the Mesozoic was indeed the Age of Reptiles—not because they were the dominant life forms, except in sheer bulk or size, but because they perhaps reached their apex of diversity as well as of size during this era.

Mass Deaths and Mass Extinctions

When it comes to cataclysmic destruction we must not suppose that survivors get off scot free while other species, other taxa, are extinguished. A score of families and some thousands of species of graptolites, for instance, were wiped out in the cataclysm that terminated the Ordovician period, while only one family survived, unless we accept that the rare present-day Pterobranchia—*Rhabdopleura* and *Cephalodiscus* (Figure 3.15)—are descendants of early primitive graptolites, as some authorities suggest. But only a single genus of that family—the monograptids—and a single species of that genus escaped the holocaust. Of that single species it seems probable that only a few individuals, making up a nuclear

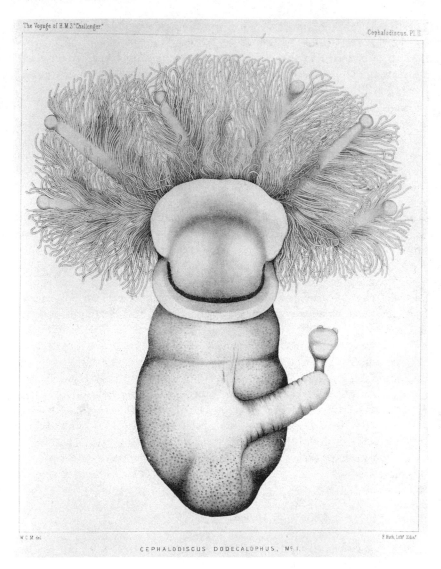

CEPHALODISCUS DODECALOPHUS, M^cI.

Fig. 3.15. Cephalodiscus dodecalophus, a pterobranch, which may be descended from the graptolites. Reprinted from the *Challenger Reports, Zoology*, vol. 20 (1887).

breeding colony, actually lived through the cataclysm; almost all individuals of even that surviving species were eliminated. It may be that a few individuals of other species also escaped destruction, but not enough to form a breeding nucleus. In that case such a species might have "survived" the catastrophe for a single generation, in the sense that some individuals (desperately seeking mates!) continued to live until they died of old age.

The same kind of thing is seen among the ammonites. Twenty thousand species swam in the seas of the Triassic period. Only two are known to have survived the catastrophe that terminated that period; these two diversified during the Jurassic and Cretaceous periods until many thousands of new species flourished, only to be extinguished entirely by the Cretaceous-Tertiary Boundary event.

At the time of the Cretaceous-Tertiary Boundary event it seems likely that *all* individual adult crocodiles and mosasaurs were killed. Since the mosasaurs, so superficially similar to crocodiles in so many ways, bore their young alive, that was the end of them. Some few crocodile eggs, however, safely buried in the sand of the riverbank, well above flood level and not due to hatch for several months, would have survived to make a new breeding population, an orphan population that never knew their parents. Turtle eggs too probably survived by the same means. Dinosaur eggs, by contrast, were laid on the surface and did not have the protection of the enshrouding sand, as did the eggs of crocodiles and turtles. Besides, there is some evidence of parental care of their young by dinosaurs. Thus, if any eggs survived the Cretaceous-Tertiary Boundary event, the newly hatched dinosaurs might not have thrived as orphans.

In general, very few individuals survived a cataclysm such as the Cretaceous-Tertiary Boundary event, probably less than one in 10,000 individuals of any surviving species. But 15 percent of species survived, in the sense that a breeding population of individuals survived. Twenty-five percent of genera and 50 percent of families lived through the event, while very few orders were actually extinguished, including both orders of dinosaurs and four other orders of reptiles, as well as the ammonites and belemnites. For families and higher taxa this rate of extinction has been established by actual counts of surviving taxa. For lower taxa—genera and species—only samples have actually been counted, and from these, several authorities have made estimates of survival and extinction.

Orders might be reduced to a single family, families to a single genus, genera to a single species, and species to a small nuclear breeding population. And to a large extent that is what "survival" usually came down to—a nuclear breeding population. That familiar children's pet the golden hamster provides an example of such survival. As best we know, every golden hamster now alive is descended from a single pregnant female trapped in Syria early this century. No more were ever found. All that exist in the wild today—and there are numerous thriving wild populations—can be traced back to escaped pets. The nuclear breeding population of this species, which had survived whatever catastrophe killed most of them, was a single pregnant female. Similarly, little more than one or two animals is what was left of thriving populations of a whole range of species after the Cretaceous-Tertiary Boundary event.

The fossil record is notoriously incomplete; soft-bodied creatures are only fossilized under quite extraordinary conditions, and even hard parts may be only rarely fossilized. Coccoliths and foraminiferan shells are the great exception; these remains are almost always fossilized if their owners are not eaten first. At the other extreme is the preservation of soft-bodied forms in the mid-Cambrian Burgess Shale of British Columbia, where submarine mudslides resulted in the preservation of a wide range of animals that are otherwise unknown in the fossil record. In general though, we have almost no fossil history of jellyfish (Figure 3.16) or of sea anemones, naked algae, or lampreys, all of which have no easily fossilized hard parts. Our counts of extinctions are thus quite biased from the nature of the record.

Within this limitation, though, it is clear that about half of the known families of animals and plants were exterminated at the Cretaceous-Tertiary Boundary, about 75 percent of genera, and 85 percent of species. David Raup of the University of Chicago has made an actual count of families; the figures for genera and species are estimates. Extinctions of higher-ranking taxa were relatively rare but included the six orders of reptiles mentioned above as well as the ammonites, the belemnites, and several orders of plants.

The mass deaths, as opposed to the mass extinctions, at this era must have been far greater. It only takes a surviving breeding pair of a species to prevent extinction, so even species that were not exterminated must have suffered extreme destruction of individuals. I have calculated that at the very least, 99.99 percent of *all* individuals of *every* species of animal and plant were killed in the Cretaceous-Tertiary Boundary event, leaving a minimal surviving breeding population in the remaining 0.01 percent (or less) of some species. For emphasis let me present in tabular form the data for deaths and extinctions at the Cretaceous-Tertiary Boundary.

Individuals	99.99% (at least)
Species	85%
Genera	75%
Families	50%
Higher taxa	a handful

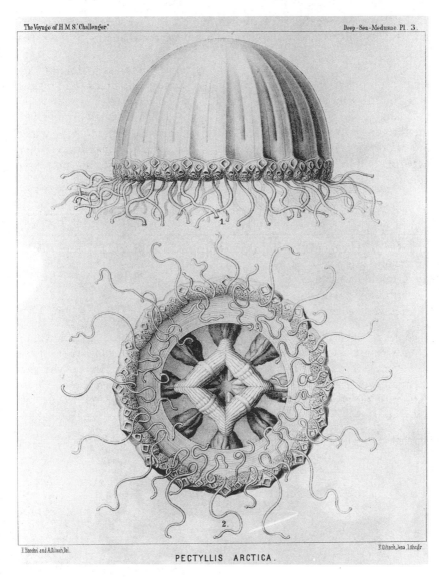

PECTYLLIS ARCTICA.

Fig. 3.16. Pectyllis arctica, a jellyfish (medusa). Soft-bodied creatures like this are rarely fossilized, though specimens are known from the mid-Cambrian Burgess Shale. Reprinted from the *Challenger Reports, Zoology*, vol. 4 (1882).

We can only assume that these figures, which are rather approximate, apply also to soft-bodied groups, which have left no fossils or very few of them.

There have been at least four more comparable mass extinctions (as well as a number of smaller ones) during Phanerozoic times. The biggest was undoubtedly the event that terminated the Permian, and with it the Paleozoic. This terminal Paleozoic event, which is associated in time with the formation by vulcanism of the Siberian Traps, appears to have wiped out about 95 percent of species living at that time. The five biggest Phanerozoic extinctions for which we have reliable estimates are presented below in tabular form.

| | Date | Extinctions (Pct.) | |
Era	(megayears)	Genera	Families
Terminal Ordovician	435	50%	25%
Devonian	357	60	25
Terminal Permian	250	80	60
Terminal Triassic	198	55	25
Cretaceous-Tertiary	65	75	50

Will our own era, marked as it is by the formation of the Hawaiian Islands (quite comparable in mass of lava to the Deccan Traps of the Cretaceous-Tertiary Boundary event and to the Siberian Traps of the final Permian event), end in a similar cataclysm? Certainly the current rate of extinction of species is also comparable to these two events, but that is our own doing, not the effect of cataclysmic events.

4

The Cretaceous-Tertiary
Boundary Around the World

Immediately [they] were annihilated, destroyed, broken up,
and killed. A flood was brought about by the Heart of
Heaven; a great flood was formed, . . . the face of the Earth
was darkened and a black rain began to fall, by day and by
night.

—*Popol Vuh* (the sacred book of the Quiché Maya)

Rocks are nowhere stable, shifting
sideways and up and down, eroding away and redepositing farther downstream. This is true even on a historical time scale; how much more then
on a geological scale! Disastrous floods and rock movements were well
known to earlier peoples. The Quiché Maya of Guatemala, inhabitants of
lowlands, record several such events in their mythology. In the *Popol Vuh*
(as translated by D. Goetz and S. G. Morley) they record the disastrous
attempts of the Forefathers to create man, each ending in such a disaster.
The first three attempts were displeasing to the Heart of Heaven, who
destroyed them by fire and flood. The excerpt that heads this chapter is
from the account of the third creation; the survivors of this cataclysm were
the monkeys.

To turn to historical events: in the Bay of Naples, Italy, we find Roman
remains of a port, Puteoli (now Pozzuoli, where Sophia Loren was born)
with a gateway to a customs house standing on the water's edge. Three
meters above the present base of the pillars of this gateway the stones have
been bored by marine organisms. In other words, the pillars, built on land
and framing a gateway originally set back some way from the sea, have
been immersed in the sea up to this 3-meter mark; and now they are above

sea level once again. It is not that the sea has risen several meters and then fallen again, for a hundred kilometers up the coast we can see nothing like it. Rather, the level of the land in this immediate neighborhood has sunk and then risen again over the last 1,900 years.

Rock movements are often accompanied by heating. Mount Vesuvius forms one side of the Bay of Naples, and Pozzuoli, across the bay, is on the edge of a less active volcanic field, the Campi Phlaegrei (or Campi Flegrei, depending on whether you prefer the old Roman name or the modern Italian). Heat and pressure, especially when acting together, are destructive of fossils—and of biochemical traces, which have been important in my research.

Rocks may be warped and thrust up into great folds: the "nappes" of the Swiss Alps, the Rocky Mountains, the Andes, and the Himalayas. In such regions it is often difficult detective work to decide which rocks are later and which earlier, which were originally higher and which lower. And such regions of intensive folding have almost inevitably destroyed fossils (though regions of less intensive folding have sometimes yielded excellent fossils). It is of little use to look in the Himalayas for fossil dinosaurs or in the Alps for fossil ammonites. The upthrusts and folds of these mountains have been brought about by the drifting and collision of continents and of the great tectonic plates that make up the crust of the world. The northward movement of the Indian subcontinent and its collision with the mass of Asia has resulted in the formation of the Himalayas—of Everest and K2, Kanchenjunga, and the Karakoram (Figure 4.1). No, these are not the places to look for the Cretaceous-Tertiary Boundary.

Most rocks, both sedimentary and igneous, are formed under the sea, and many are swallowed up in the downward turn (the subduction, as it is called) of the moving plates of the earth's crust. The sea floor off California is being swallowed up as it moves under the land and pushes up the Sierras and the Rockies. But not all of the rocks newly formed on the sea floor are swallowed up in this way. Some are elevated by the mountain-forming processes and become dry land. As long as they have been minimally folded (or better still, not folded at all) and have not been too much compressed or heated, then we may find fossils. The gently uplifted beds that fringe the great mountain belts are where we may usefully search for fossiliferous rocks.

Cretaceous-Tertiary Boundary Sites

In Chapter 2 I have described how the Badlands of Alberta have been lifted almost a kilometer above where they were formed, lifted without

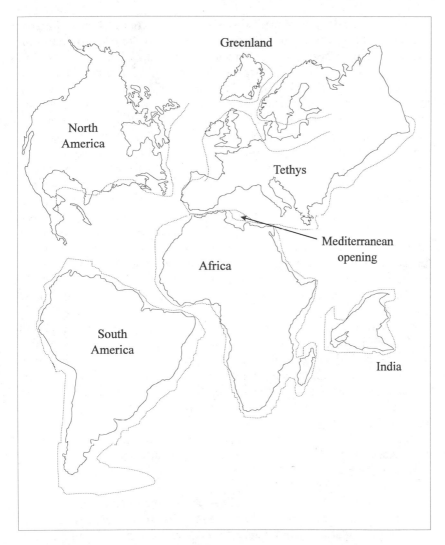

Fig. 4.1. Map of the tectonic plates around the newly forming Atlantic Ocean at the end of the Cretaceous era. The Atlantic had formed but was much smaller than today. Africa was just starting to break away from Tethys. Greenland had only recently broken away from the boreal land mass of Scandinavia and Siberia, itself separated from Tethys by an inland sea running through Denmark. North America had started its counterclockwise rotation but was still 30 degrees away from its present orientation. India had broken away from Antarctica and was steadily plowing its way northward toward its eventual collision with Asia, rotating clockwise as it did so. (In this map the European land mass is cut off along the current geographical boundary with Asia.)

deformation or even significant tilting—the beds are only about one degree off the horizontal even today. Since they have never been folded, compressed, tilted, or heated, they are ideal sites for searching for fossils.

It is not, however, enough that the rocks should have been lifted undisturbed in this way if we are to find the fossiliferous beds that interest us. In addition, something must have cut through the beds to expose a section of rocks. In the Badlands of Alberta, the Red Deer River has cut a canyon several hundred feet deep through the rocks, slicing through the Cretaceous-Tertiary Boundary and thus exposing the section we want.

Other exposures are found in the cliffs of the seacoast, for instance at Zumaya in Spain or Biarritz in France. At other sites, quarrying has exposed the rocks we are looking for. Also, drilling for oil may bring up cores that cut through the rock strata we wish to study. Some geologists use oil-drilling equipment not with the intention of looking for oil but simply to find the rock strata that interest them. As I write this, a group of Canadian geologists are planning just such a drilling on the prairies and have asked me to collaborate with them.

The unique features of the rock sections exposed in the Red Deer Valley attracted me to this site in the first place, because I felt that there I could test some of my burgeoning theoretical ideas. Other sites have attracted other scientists because of their unique features, their good documentation, or simply their accessibility. The Gubbio site in Italy, for instance, was first found by an Italian scientist, Isabella Premoli Silva, to whom it was accessible. It was also the site of the first studies of the "iridium anomaly" by Luis and Walter Alvarez and their colleagues because the area had been well surveyed by Silva.

Now there are a hundred or more sites where the Cretaceous-Tertiary Boundary is exposed, a score of which are readily accessible. A few of these are composed of rocks laid down on land; in others the rocks are marine in origin (Figure 4.2).

Sixty-five million years ago North America contained a great shallow sea, running north and south where the prairies are now located. Eastern Alberta was under this sea; western Alberta was land at that time, sloping down from the Rocky Mountains, which were already beginning to form. The mountain slopes were steadily eroding, depositing silt in the lowlands and out at sea. These lowlands have since been elevated almost a mile and form the western part of the prairies. The rocks formed at the same time, offshore in the shallow sea, are the base of the eastern part of the prairies. We do not, however, have any good section of the Cretaceous-Tertiary Boundary in the marine rocks of this area. Only by drilling might we find the boundary rocks.

Fig. 4.2. Map of some of the better-known Cretaceous-Tertiary Boundary sites.

At that time these lowlands of the western North American continent, like the Mountains of Morne, "swept down to the sea," extending from Alberta into Montana, Wyoming, Colorado, and finally into Texas. Later Cretaceous and early Tertiary rocks are exposed in all these states, with dinosaur remains and occasional glimpses of the Cretaceous-Tertiary Boundary. Oil drilling in Texas has also revealed more of the Cretaceous-Tertiary Boundary in that state. A little farther east, marine rocks of the same age are exposed at locations in the Dakotas, Kansas, and Arizona, but these, of course, have no dinosaur remains.

In fact, the known sites represent a quite select group of habitats from the end of the Cretaceous and the beginning of the Tertiary, sites where deposition was taking place. By their very nature, erosional sites, whether on a steep hillside or a coastal cliff, cannot form new rock; rather, old rock is continually being eroded away. Neither can the deep sea afford sediments preserved for any great length of time; these rocks are being subducted under continental rocks, and there are no rock cuts where older sediments could be seen. We have some record of the Cretaceous-Tertiary Boundary in cores taken by drilling ships in the deep sea, but little has turned up in the way of fossils, except for microfossils. Deep-sea drilling, however, has revealed one thing of importance: the Cretaceous-Tertiary

Boundary event does not coincide in time with any of the periodic magnetic reversals but rather occurred almost halfway between two of them. In other words, whatever caused the magnetic reversals could not have caused the Cretaceous-Tertiary Boundary event. In fact this event took place during the magnetic chron (as it is called) number 29R (R for reversed magnetic field, when the north magnetic pole was near the south geographical pole). Apparently, the Cretaceous-Tertiary Boundary event was quite dissociated from any magnetic reversal.

Flora and Fauna at the Cretaceous-Tertiary Boundary

Terrestrial Vegetation of North America

At the end of the Cretaceous, flowering plants were abundant but largely confined to stream margins and to the undergrowth of forests. The countryside was much warmer than today, so that tropical-type vegetation extended 20 degrees farther north and temperate vegetation 25 degrees farther north than nowadays. The coastline of Alberta (what is now Saskatchewan was at that time an inland sea) supported an almost subtropical forest, and the average temperature throughout the year was similar to that of present-day Florida.

From the fossil record we know of three main vegetation types from the late Cretaceous. From Georgia to Texas, the land was covered by a tropical evergreen woodland consisting largely of conifers that had broad leaves instead of the needles of most of our conifers today. Flowering plants, including shrubs, were found under these trees and along the edges of streams. Cycads and their relatives were almost absent from this region, where the soils were well drained, the trees widely spaced, and their stature limited by a low supply of soil nutrients. The climate seems to have been equable all the year round, with little sign of separate seasons.

The northern western interior and the Pacific coast supported sub-humid broad-leaved evergreen forests that existed under cooler and wetter conditions, with more definite seasons. The trees were more closely spaced and festooned with vines and lianas. About two-thirds of the species were flowering plants, and cycads and ginkgos were abundant.

Arctic Alaska supported a broad-leaved deciduous forest in general aspect not unlike the woodlands of Virginia at the present day. The forest was dominated by angiosperms, like modern broad-leaved forest trees, and most of the conifers too shed their leaves, unlike modern pines and cedars. Large leaf size, narrow tree rings, and abundant coal formation indicate that this forest grew under conditions of abundant moisture. Like

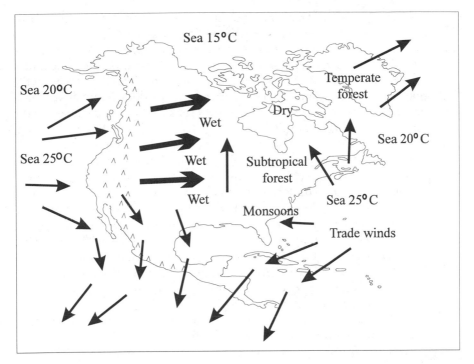

Fig. 4.3. The climate of North America at the end of the Cretaceous. The arrows indicate prevailing winds, with the width indicating mean speeds. The interior seaway was crossed by savage westerly gales.

the modern muskeg or the swamp cypress of Lousiana, many of these trees grew with their feet in the water, in swampy or marshy conditions. In the lowlands of the Alberta coast there were relatively few trees standing in the marshy land, which was frequently flooded by runoff from the mountains behind (Figure 4.3).

The widespread destruction of the Cretaceous ecosystems produced profound vegetational changes at the Cretaceous-Tertiary Boundary. Fossils of spores, pollen grains, and leaves show a mass-kill event at this epoch, which was followed by a pattern of recovery. After this initial recovery the early Tertiary vegetation differed from that of the Cretaceous in two fundamental respects. First, the forests were much more strongly stratified, showing clearer layering into upper-story trees, understory small trees, shrubs, and ground vegetation below. Second, the temperate forests were almost wiped out, so that boreal forests abutted directly on the subtropical.

Let me put some numbers to this. In the latest Cretaceous, from New

Mexico to southern Canada, 70 to 80 percent of the microfossils consist of the pollen of flowering plants (including trees), and species unique to the Cretaceous persist right up to the Cretaceous-Tertiary Boundary. The boundary claystone itself has no pollen or other microfossils. Immediately above the Cretaceous-Tertiary Boundary, the absolute abundance of the pollen of flowering plants declines abruptly, as does the relative abundance, and the proportion of fern spores rises to 70 to 100 percent of the microfossils. This "fern spike," as it is generally called, is known throughout North America, from many different kinds of rocks and is found only in the few centimeters above the Cretaceous-Tertiary Boundary. In each locality only a few species of ferns are present, being different in different places. In most localities this fern spike is present in only the lowest 15 centimeters of Tertiary rock, immediately overlying the Cretaceous-Tertiary Boundary claystone, and is succeeded by a decline in relative abundance of fern spores and a concomitant rise in pollen. This is exactly what we observe today in regions devastated by volcanic eruptions. At the Mexican volcano El Chichón, for example, after the recent eruption the lava and mudflows were colonized first by ferns and mosses; only today are these being replaced by other vegetation.

In the more northerly climate of Alberta, the most notable effect of the Cretaceous-Tertiary Boundary event on the vegetation is the impoverishment of the species diversity, though here too there is a fern spike. Below the Cretaceous-Tertiary Boundary, and right up to it, the range of species is rather diverse and quite rich. Above the boundary the flora is dominated by a very few species of ferns and a few flowering plants, but within about 10 to 20 centimeters the vegetation becomes dominated by ferns and conifers, mostly cypresses and junipers. There was clearly widespread frequent flooding, giving rise to coal formation, as we have seen at the Red Deer Valley section.

One obvious feature of the survivors is that the northern vegetation fared better than the more tropical forests, probably because northern plants are characterized by dormancy mechanisms—overwintering seeds and spores, for example—that enabled them to weather the storm, even if adult trees were wiped out. The recolonization of Mount Saint Helens at the present day is reminiscent of what we can infer from the succession of spores and pollen grains above the northern exposures of the Cretaceous-Tertiary Boundary. Given this, the evergreen forests of southern North America were far more severely affected by the Cretaceous-Tertiary Boundary event than the northern.

The record of fossil plants suggests that the climate changed abruptly after the Cretaceous-Tertiary Boundary event, not so much in a drop in average temperature at any one place, though this certainly occurred, at least as a brief episode, but rather in greater seasonal extremes. Subtropical areas in which seasonal changes had been unknown in the Cretaceous were subjected to cold winters and warm summers, even to occasional freezes in winter. As a result, deciduous trees, which can shed their leaves in winter, became common, replacing the evergreens.

During the early Paleocene (the earliest part of the Tertiary), rainfall increased throughout North America, and this is evident in the coal formation already noted. More than half the trees possessed leaves with "drip tips," long points that enable them to shed excess rainfall.

Aquatic Vegetation

We have no such complete record of changes in aquatic vegetation, having few leaf impressions of freshwater plants and almost no remains of kelps or seaweeds. By contrast with this paucity of remains of larger plants, we have abundant microfossils, chiefly single-celled algae, the phytoplankton. As at the present day, phytoplankton accounted for most of the global oxygen production and for the primary productivity of the earth, far more than did the larger vegetation. Today phytoplankton produce perhaps 90 percent of the oxygen of the atmosphere.

In the late Cretaceous, lakes were dominated largely by blue-green algae (also known as blue-green bacteria) and by naked flagellates. The tropical and temperate seas were dominated by coccolithophorids, those green algae armored with tiny chalky plates. Arctic waters were dominated by diatoms, algae with siliceous skeletons specialized for arctic existence by the ability to produce resting spores that could survive the arctic night, and to a lesser extent by dinoflagellates, which also produce resting spores.

The Cretaceous-Tertiary Boundary event led to the demise of the coccolithophorids, and chalk ceased to be formed. Blue-green algae too diminished in importance, both in the lakes and in the seas, so that today these creatures are regarded as markers of polluted waters, especially of polluted waters rich in phosphates. In their place there was a rapid domination of the territory by diatoms, and a previously minor group, the dinoflagellates, came to dominance in tropical and temperate waters, both fresh and salt. Most marine deposits from the very earliest part of the Paleocene are dominated by diatoms, but soon dinoflagellates increased in number in all but the coldest waters until within a million years they had reached the dominance they have today.

Marine Fauna

During late Cretaceous times the pelagic zone of the sea was dominated by ammonites and the much smaller single-celled foraminifera (Figure 3.2). Both had calcareous skeletons, and both disappeared at the Cretaceous-Tertiary Boundary. Ammonites disappeared totally, as did their close relatives the belemnites, from which squid are descended; all pelagic foraminifera too were extinguished, but the groups survived in the form of benthic species inhabiting the sea bottom near coral reefs. Planktonic foraminifera of the present day are descendants not of Cretaceous planktonic forms but of these bottom-dwelling coral-reef species.

After the Cretaceous-Tertiary Boundary event, the place of both ammonites and Foraminifera in the plankton was taken over largely by planktonic crustaceans, chiefly the copepods, which today form about 90 percent of the total biomass of animal life on the earth. In fact, although biological evidence suggests that Copepoda may have originated either just before or just after the great Permian extinction, some 250 million years ago, they are almost unknown in the fossil record before the Tertiary. Today copepods are the main grazing animals in all the oceans, feeding on the planktonic algae—the dinoflagellates and diatoms, and other less abundant groups. Some relatives of the ammonites survived: the carnivorous squid, whose skeleton has become totally internal and lacking in any calcium, and five species of nautiloids. But it was the crustaceans, and especially the copepods, that inherited the seas. Like the squid, copepods lack calcium carbonate as a stiffening material in their skeletons, relying on chitin and tanned protein instead. Their relatives the Decapoda (crabs, lobsters, shrimps, and their allies) by contrast possess calcified shells (as anyone knows who has eaten a lobster). As with most taxa, nearly all decapod genera and many families were wiped out, and most present-day taxa of crabs, shrimps, and lobsters have evolved since the Cretaceous-Tertiary Boundary from those few species that survived. Similarly, barnacles, those most abundant denizens of rocky shores and the hulls of ships, were decimated at the Cretaceous-Tertiary Boundary, as were ostracods and other groups of crustaceans that were notable in the Cretaceous period.

Like other marine groups with a calcium carbonate skeleton, the echinoderms were almost wiped out. Ninety percent of families of sea urchins were extinguished, and most of those alive today have evolved since the Cretaceous-Tertiary Boundary event. Starfish too barely survived the cataclysm, but the deep-water crinoids (Figure 4.4), or sea lilies, were relatively little affected, though the stalked forms, which were more abundant before the Cretaceous-Tertiary Boundary, were largely replaced by the stalkless forms (Figure 4.5).

Sponges come in three main varieties: those with a spongy skeleton, those with a calcareous one (Figure 4.6), and those with a skeleton of silica (Figure 4.7). During the Mesozoic the calcareous sponges developed the

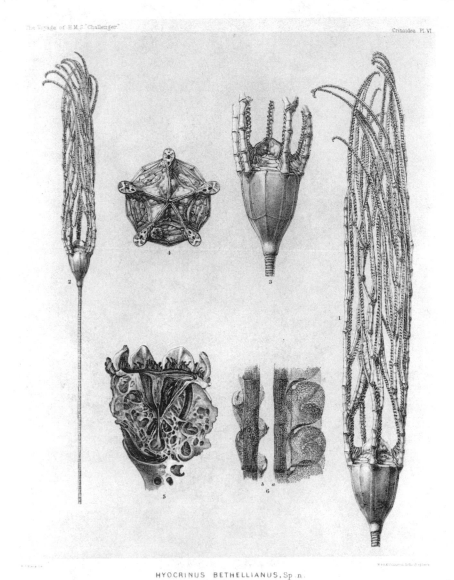

Fig. 4.4. Hyocrinus bethellianus, a stalked crinoid, or sea lily. Stalked crinoids were common before the Cretaceous-Tertiary Boundary. Reprinted from the *Challenger Reports, Zoology,* vol. 11 (1884).

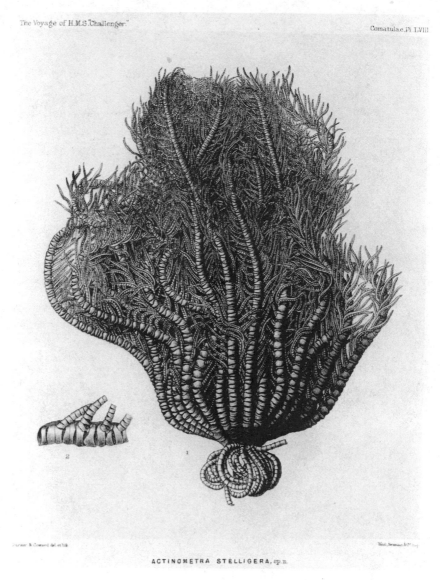

ACTINOMETRA STELLIGERA, sp.n.

Fig. 4.5. Actinometra stelligera, an unstalked crinoid. Today unstalked crinoids like this one are much more abundant than stalked forms. Reprinted from the *Challenger Reports, Zoology*, vol. 26 (1888).

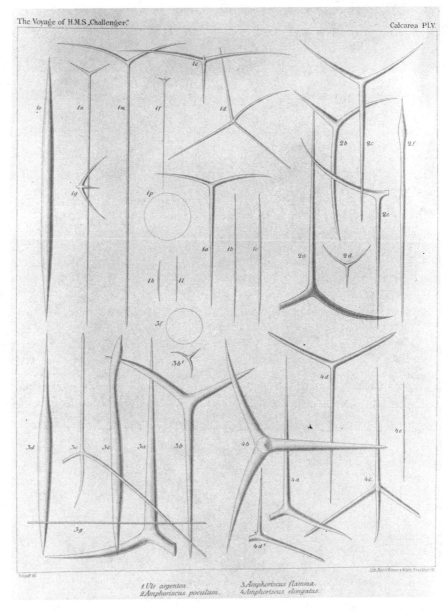

Fig. 4.6. Spicules from a calcareous sponge. Sponges of this type once formed reefs, like coral reefs, before the Cretaceous-Tertiary Boundary. Reprinted from the *Challenger Reports, Zoology*, vol. 8 (1883).

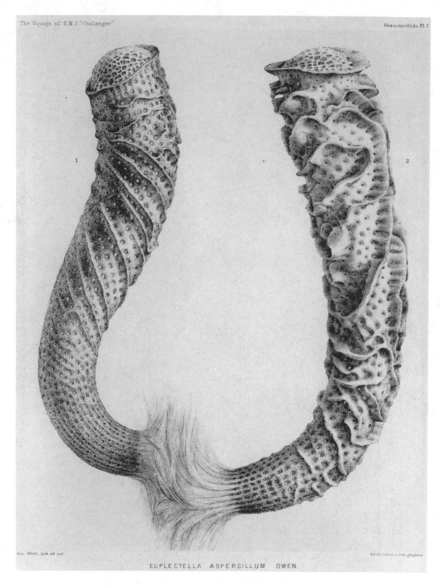

The Voyage of H.M.S."Challenger"   Hexactinellida Pl. I

EUPLECTELLA ASPERGILLUM OWEN.

Fig. 4.7. Euplectella aspergillum, Venus's flower basket, a sponge with a siliceous skeleton. Reprinted from the *Challenger Reports, Zoology*, vol. 21 (1887).

ability to form reefs, rather like coral reefs, but lying deeper in the ocean, below normal wave base but within the range of storm waves. By the Cretaceous these sponge reefs were smaller and included many siliceous sponges as well. The reefs were totally destroyed at the Cretaceous-Tertiary Boundary, and all subsequent sponges, as today, have not formed reefs.

We have no records of the fate of the tunicates (Figure 4.8), those relatives of the vertebrates that must have evolved early in the Paleozoic. They have no hard parts and are almost unknown from the fossil record. Today there are about 20,000 species in about twenty families around the globe.

In the immediate aftermath of the Cretaceous-Tertiary Boundary event, there is little evidence of any planktonic animals at all. The marine beds of the early Tertiary, which show abundant phytoplankton, chiefly diatoms, are largely devoid of zooplankton. The same, of course, is largely true of the terrestrial rocks; there are few faunistic remains to be found during the fern spike of the lowermost Tertiary. It is only on fossil coral reefs that we find significant animal remains: fish, with their calcium phosphate skeletons; *Sepia* (cuttlefish); worms; snails and clams; and the corals themselves. But not just any corals. Among all these taxa, species diversity was enormously reduced, and two families of corals (the Microsolenidae and the Amphiastreidae) disappeared entirely. Limestone, which earlier had been largely the fossil remains of coral reefs, now became more a sign of chemical deposition. The formation of extensive coral reefs largely ceased. Instead of being taken up by marine animals and being largely used up in skeleton formation, there was now an excess of calcium carbonate in the sea, which led to chemical deposition. It was not until the some ten million years had passed that this deposition ceased, and calcium became once more in short supply. The famous section of the Cretaceous-Tertiary Boundary at Gubbio in Italy represents a fossil coral reef, while the Stevns Klint site in Denmark is poorly developed coral below and largely chemically precipitated limestone above the Cretaceous-Tertiary Boundary.

The general impression one receives from the study of marine rocks across the Cretaceous-Tertiary Boundary is that pelagic animals were almost entirely wiped out and that the only survivors were those that inhabited coral reefs. Marine shales, for example, often carry heavy deposits of microfossils below the Cretaceous-Tertiary Boundary, while above it are found little but terrestrial fern spores and pollen grains blown from the land. Plesiosaurs and ichthyosaurs, those giant marine reptiles, disappear entirely from the fossil record (though ichthyosaurs may already have been extinct by the end of the Cretaceous), leaving only turtles to represent

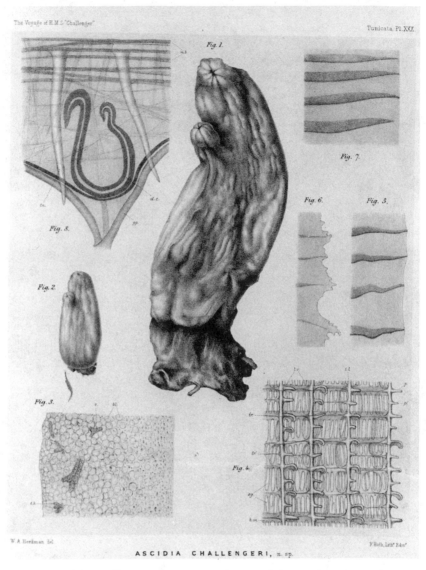

Fig. 4.8. Ascidia challengeri, a tunicate. Relatives of the vertebrates, tunicates are almost absent from the fossil record, yet evolved early in the Paleozoic and have survived to the present day. Reprinted from the *Challenger Reports, Zoology*, vol. 6 (1887).

the reptile class in the sea. Whales, of course, had not yet evolved, nor had seals, and there were no marine mammals as best we can tell.

Terrestrial Faunas

Though as scientists we know far more about aquatic animals and plants than about terrestrial groups, it is these, especially the dinosaurs, whose extinction has attracted the most attention. Unlike the seas, the land has very few single-celled animals, and they have never been prominent in the fossil record. Even soil organisms, which include many species of protozoans, are not well represented as fossils, and neither are terrestrial worms, which lack the easily preserved scales and setae (or bristles) that characterize their marine relatives (Figure 4.9). The terrestrial faunistic fossil record is dominated by larger forms.

Today the insects dominate terrestrial faunas, both in numbers and in number of species and even in biomass. The earliest insects are known from the Devonian period, and by the Carboniferous period we already have fossil insects surprisingly like their present-day counterparts. The Carboniferous forests supported insect-eating dragonflies, omnivorous cockroaches, and many other groups of insects. No single known order of insects was wiped out in its entirety at the Cretaceous-Tertiary Boundary, though eight had been extinguished at the earlier Permian-Triassic event. Mesozoic insects are not too well known (they are not easily preserved as fossils), and late Cretaceous fossil insects are particularly scarce. In consequence, we have little idea of how many species or genera were wiped out at the Cretaceous-Tertiary Boundary. We do, however, know that at least six orders of insects first evolved or radiated from other orders soon after that event. These include the important Lepidoptera (butterflies and moths, including such major pests of agriculture as the cutworm and the tobacco hornworm) and termites. Fleas too are unknown as fossils before the Cretaceous-Tertiary Boundary and presumably evolved with their hosts, the mammals.

Most of the insects we know from the Mesozoic era were preserved in amber, and it is not too farfetched to suppose that dinosaur blood and dinosaur DNA could be recovered from the guts of biting flies preserved in this way. This of course is the central plot feature of the scenario of Michael Crichton's novel *Jurassic Park*. Unfortunately for the verisimilitude of that novel, mosquitoes seem to have evolved after the Cretaceous-Tertiary Boundary as parasites on mammals. We know of none from the Mesozoic, though other (nonbiting) flies are common in amber; biting flies are common in more recent amber.

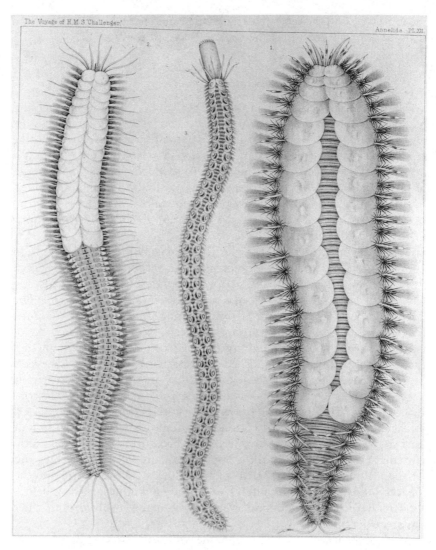

Fig. 4.9. Hermadion, Polyeunoa, and *Polynoe,* marine polychaete worms. The scales and setae of worms of this type are abundant in the fossil record. Reprinted from the *Challenger Reports, Zoology,* vol. 12 (1885).

The Cretaceous-Tertiary Boundary event clearly affected all groups of animals and plants in all "the round Earth's imagined corners." It was not a local event, but one that demolished almost all living things, perhaps more than 99.99 percent of individuals, causing the extinction of 85 percent of known species all over the world.

The Solar System, Vortices, and Comets

It is essential to have some grasp of the evolution of the solar system in order to begin to understand the physical nature and magnitude of the Cretaceous-Tertiary Boundary event, for nothing less than an extraterrestrial happening could cause the devastation we see in the fossil record.

The Solar System: Structure and Origins

The solar system consists of several zones around the sun. First comes the inner solar system, consisting of the planets and asteroids. This is divided into two portions, the inner zone of the stony planets, together with the asteroids (in the asteroid belt), and the outer zone of the gas giants and the outer planets. Beyond the inner solar system lies an empty zone, which in my view is best regarded as a forbidden zone, and then comes the Oort Cloud of comets, comprising the outer solar system (Figure 5.1).

The components of the zone of stony planets are, successively, Mercury, Venus, Earth, Mars, and the asteroids (Figure 5.1D). There are perhaps 4,000 asteroids more than 10 kilometers across, a million that are more than 1 kilometer. The largest of these is Ceres.

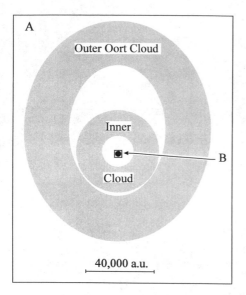

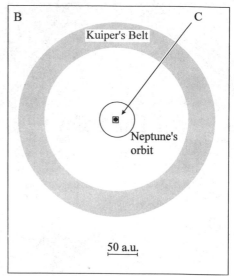

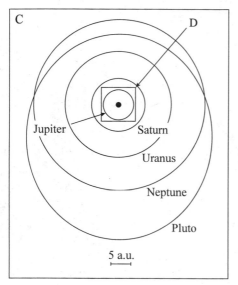

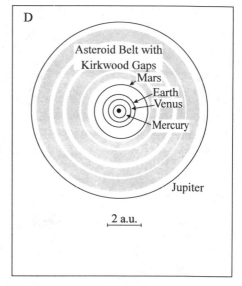

Fig. 5.1. The structure of the solar system, approximately to scale. Note the differences in the scale bars. The figure is in four parts: A, the whole solar system, including the Oort Cloud; B, Kuiper Belt; C, the gas giants; D, the stony planets and asteroid belt. These parts are nested one within another. For example, the rectangle labeled B in Figure 5.1A is enlarged to become Figure 5.1B, and so on.

The zone of the gas giants comprises Jupiter, Saturn, Neptune, and Uranus, together with Pluto, which is hardly a gas giant but orbits in this zone (Figure 5.1C). Pluto is in a highly oblique and somewhat eccentric orbit, rising at times far above the plane of the ecliptic, the plane near which all the other planets lie, and spending part of its orbit outside the orbit of Neptune and part inside. At the moment, and for many years into the future, it lies inside the orbit of Neptune. Its orbit is unstable and it cannot have been there since the beginning of the solar system, four and a half billion years ago. It was probably captured from a passing star.

A tenth planet has been suggested, to account for possible perturbations in the orbit of Neptune (and of Pluto). If it exists at all, this planet X would orbit a little beyond the outer limit of the orbit of Pluto, but recent measurements of Neptune's trajectory suggest that the supposed discrepancies lie well within the limits of observational error. Beyond Neptune and Pluto (or beyond planet X, if it exists) comes the forbidden zone and then, some 100 milliparsecs from the sun, the inner limit of the Oort Cloud of comets, containing over 10^{13} comets and extending almost halfway to the nearest star.

To give some idea of scale and of units of distance, the nearest star, Alpha Centauri, is about 1.2 parsecs away, or 4 light-years. A parsec is about 3.26 light-years, so 100 milliparsecs (abbreviated mpc) is about 4 light-months. For comparison, the earth is about 8 light-minutes from the sun. This distance, the mean distance from the center of the sun to the center of the earth, is known as an astronomical unit (abbreviated a.u.). A parsec is thus about 210,000 astronomical units, while 100 milliparsecs (the distance to the inner edge of the Oort Cloud) is about 21 thousand times the distance of the earth from the sun. I shall use both parsecs and astronomical units in this book and avoid the use of light-years as a measure of distance.

Both the asteroid belt and the Oort Cloud show gaps in distribution devoid of any celestial bodies. The gaps in the asteroid belt were first observed by the American astronomer Daniel Kirkwood in the middle of the nineteenth century and are named for him. Only a single gap is known to occur in the Oort Cloud, and that is more a matter of calculation than of observation; it divides the Oort Cloud into inner and outer regions, usually called the inner and outer clouds (Figure 5.1A).

The Kirkwood gaps have been shown to be a result of resonance chiefly with Jupiter but also with other planets. To oversimplify, the resonance with Jupiter sweeps clear of asteroids those gaps whose distance from the sun bears a simple arithmetic relationship to Jupiter's distance from the sun. Any asteroids orbiting in these gaps are subject to perturbation at each pass of the planet and gradually move out to other orbits in the asteroid belt. To some extent Saturn too may also influence the Kirkwood gaps.

In addition to the Oort Cloud there is a second belt of comets just outside Neptune. I shall show later that this belt, known as the Kuiper Belt, is unstable, but is repeatedly replenished from the Oort Cloud (Figure 5.1B). It is the source of short-period comets, such as comet Halley, which return regularly to the inner solar system from this belt, just beyond Neptune.

There is a steady gradient in chemical composition as we move away from the sun. The innermost stony planets and the asteroids are all rich in metals. In the larger bodies of the innermost solar system, including the larger asteroids, the metals have sunk toward the center to form a metallic core during a period in which the bodies were molten. The lighter silicate rocks floated as a slag on top of the metallic core, later hardening to form the crustal rocks and mantle. Some asteroids have been broken apart at a subsequent date by collisions, so that some at the present day lack metals, while others are essentially metallic—the former representing bits of crust and mantle from the original body, and the latter, parts of the metallic core. In contrast Jupiter, and perhaps the other gas giants, seems to have a core of metallic solid hydrogen and to be relatively devoid of the heavier metals. There is even a possibility that part of the core of Jupiter consists of diamond, a suggestion seized on by Arthur C. Clarke in his novel *2061*.

There is also a gradient in oxygen abundance, with the innermost planets being richer in this element. This is evinced most clearly in the formation of silicate rocks, which are abundant throughout this zone of stony planets and asteroids. In these, excess oxygen has been mopped up by reactions with silicon to form silica and silicates. Carbon is less abundant than oxygen, so the commonest form of carbon (in the absence of lifeforms, which upset the natural balance) is carbon dioxide. Likewise, almost all hydrogen is oxidized to water.

In the region of the gas giants oxygen is less abundant, so that in Jupiter, for instance, there is much free hydrogen and carbon. The moons of these planets may contain not carbon dioxide but methane, formed by the reaction of carbon with excess hydrogen. Still farther out, in the Oort Cloud, oxygen is even less abundant compared with carbon, so that methane, organic compounds, and free carbon are abundant in comets. The free carbon may take the form of graphite, amorphous carbon, or diamonds; perhaps 10 percent of the free carbon in comets is in the form of diamonds. Some carbon has also reacted with silicon to form silicon carbide, a compound known on earth chiefly as a component of manmade artifacts.

A further difference between the zone of the stony planets and that of the Oort Cloud lies in the relative abundance of the stable isotopes of carbon. The comets have the primordial ratio of these two isotopes, roughly 42 times as much carbon 12 as carbon 13, the same ratio as grains of diamond and of graphite in interstellar space. The earth, and the stony planets in general, have more than twice the relative amount of carbon 12, about 91 times as much as carbon 13. There have been no sensible suggestions made so far to account for the relative enrichment of carbon 12 in the stony planets nor of heavy metals, though in general terms a gradient from the sun to interstellar space seems not unreasonable.

These differences in composition help us to identify the source of meteorites, all of which come from outside the earth but from within the solar system. Stony objects have come from the innermost zone of the solar system, mostly from the asteroids, but a few from the moon or Mars. Metallic objects seem to originate as parts of the cores of asteroids that have been smashed apart in collisions. One example would be the bolide whose impact 1.82 billion years ago created the Sudbury nickel deposits in northern Ontario (and probably led to a mass extinction far greater than any we know from the fossil record, since its size was much greater than more recent impactors). Another would be the iron bolide that created the Barringer Crater in Arizona some 30,000 to 40,000 years ago. Finally, meteorites rich in free carbon and organic compounds, with a carbon-12-to-carbon-13 ratio near 42:1 and containing silicon carbide and diamonds, seem to be fragments of comets. These last are called carbonaceous meteorites.

The solar system is not contemporaneous with the universe, or even with our galaxy, the Milky Way, which was already at least ten billion years old when the sun and solar system were formed. The universe began in the big bang (not to be confused with the "little" bang, which is the real subject of this book), and perhaps a billion years later the Milky Way was formed. Giant clouds of gas condensed to form star clusters. The stars ran through their normal life and died, some of them spectacularly as supernovae or novae. These first stars were initially composed of just hydrogen and helium, the only two elements formed in any large amounts in the big bang; they obtain their energy by nucleosynthesis, which involves the fusion of these atomic nuclei to heavier elements, up to iron of mass 56. The death of the stars, when they had "burnt" all their nuclear fuel, released these elements into the surrounding gas, while supernova explosions produced even heavier elements. All the elements with a higher mass than iron 56—gold and platinum, uranium and plutonium—were produced in supernova explosions.

Not all or even the majority of the gas in the Milky Way condensed into star clusters in the first era of star formation. While the first stars were running through their lives, self-gravitation resulted in the formation of more giant clouds of denser gas, so that when large stars blew their tops in supernova explosions, the shock wave sometimes triggered further condensation of these clouds to form new stars, this time containing not just hydrogen and helium but also the heavier elements formed by nucleosynthesis within the earlier generation of stars and the even heavier elements formed in supernova explosions.

The sun is not even a second-generation star. From the relative amounts of heavy elements in the solar system, we can calculate that our sun is a third-generation star, since it contains much more than its fair share of these elements. It was formed about four and a half billion years ago from a gas cloud whose condensation was triggered by one or more supernova explosions.

Before I continue with the evolution of the solar system, I must introduce you to some of my ideas about vortices. I will warn you that these ideas are not conventional at all but highly idiosyncratic.

Vortices

Have you ever considered what happens when you stir a cup of coffee? You have put in the sugar, which rests on the bottom of the cup, added cream, which has not yet mixed with the coffee, and then you begin to stir. The sugar dissolves and both sugar and cream are mixed with the coffee. What has happened? The stirring was strictly horizontal, with no vertical component, but the sugar solution has been lifted from the bottom of the cup, while the cream, floating near the surface, has been distributed equally throughout the coffee. Where did the vertical movement of cream and sugar come from?

Observe the surface of the coffee. The liquid around the edge rises (you can't stir a cup properly if it is too full; it will just slop over), and the center is depressed. You have in fact created a vortex. If you stop stirring, the vortex dies down by friction with the inner surface of the cup.

This everyday observation illustrates two important facts about vortices: first, they can only originate and end at a surface, never in the body of a fluid (starting in this example at the surface of the spoon and ending by friction with the surface of the cup); and second, they are not just two-dimensional. The surface of the spoon or coffee stirrer starts the vortex; the shape or the mass of the spoon does not matter, only its surface. Vorticity, the eddy motion of the liquid, will continue until friction with a surface kills the vortex.

The second point is more subtle. Several formal proofs were offered in the nineteenth century showing that a vortex is more than two-dimensional, as agrees with common observation. As fluid moves horizontally at the surface of a vortex, it must finally move out of the plane, thus supplying the vertical component that results in the mixing of cream and sugar with the coffee. Engineers have always assumed that this implies that a real vortex is three-dimensional, but another nineteenth-century proof, a nonintuitive one this time, demonstrates that a vortex is less than three-dimensional. It is easy to think that this is a proof that a vortex is not more than three-dimensional or less than or equal to three-dimensional, and generally engineers have proceeded accordingly. Unfortunately, if you try to calculate the properties of any flow pattern that involves vortical flow (the flow of air around an aircraft wing, for instance), you will sooner or later find yourself trying to divide by zero. At this point in such calculations, it is usual to introduce an "empirical adjustment," otherwise known as a "fudge factor." This need stems from assuming that a vortex is three-dimensional. It is not. It exists simultaneously in some space that has more than two dimensions but less than three.

If you have come across fractals, then you will be aware that complex shapes can exist in a fractional number of dimensions, and this is just what governs vortices. What is the correct number of dimensions for calculating the properties of any real vortex? To answer this we must first consider how we describe the trajectory of a circulating object—not a fluid, but an object like the earth moving around the sun, or a space shuttle moving around the earth.

When you walk along a footpath you see yourself as moving straight along, but to an observer in space you are moving along a curve, the curved surface of the earth. It all depends on the frame of reference of the observer. In fact your motion, or that of the space shuttle, can be described in six components: three linear (forward/backward, sideways, up/down) and three rotational (pitch, roll, and yaw). What seems linear forward to me as I walk along a straight path appears as pitch to the observer in a space shuttle as I walk along the curve of the earth. At any one time all motion appears to consist of these three linear and three rotational moves, but which is linear and which is rotational depends on the frame of reference. These components are interchangeable and mathematically indistinguishable. Since three of them are our familiar dimensions, it is best to regard them all as dimensions, at least in mathematical terms. This is exactly what is done when computing the trajectory of the space shuttles. We perform our calculations in six dimensions, or to express it another way, we calculate in six-space.

Can we apply this formalism to stirring a cup of coffee? Certainly we can if we regard the coffee as a collection of discrete objects—molecules of water, caffeine, and sugar and droplets of cream—instead of as a fluid continuum. The trajectory of each particle can be described in six-dimensional space (usually written six-space), and statistically the combined trajectories of all the particles could be so described, a fairly useless and very laborious task. It does, however, point the way to resolving the problem of the dimensionality of the vortex—namely, how it can be more than two but less than three. Remember the role of surfaces in creating and killing vortices? A surface is two-dimensional. Mathematically it is convenient to regard a vortex as a mapping from the six-space needed to describe the trajectories onto the two-space of the controlling surfaces. This hypothetical mapping space has always been regarded by engineers as being three-dimensional—that is, possessing a dimensionality of six divided by two (from the six-space of trajectory description divided by the two-space of the control surface)—but this is incorrect. The true value is the logarithm of six divided by the logarithm of two, a value close to 2.58 (the proof of this is rather obscure and it serves no purpose to outline it here). Thus, a vortex is best regarded (for mathematical purposes) as a 2.58-dimensional entity embedded in three-space, "where length, breadth and highth are lost" indeed.

It all sounds rather odd, but when you attempt to compute, for instance, the flow of air around a golf ball, with all its dimples, you do not need any fudge factor when working in 2.58-space, and you can predict the range of a drive quite easily and quite accurately. The same applies to calculations about the solar system, if you regard the totality of planets, asteroids, comets, and other objects as a vortex, a 2.58-dimensional entity embedded in the "illimitable ocean" of three-space.

Some of the prints of the great Netherlands artist M. C. Escher (Figure 5.2) show a dimensional ambiguity, resulting from the depiction in two-space—the plane of the paper—of an imagined three-dimensional scene. We know that the scene is impossible, yet it is beautifully portrayed. The impossibilities become possible from the artifice of depicting three-space in two-space. To a mathematician, the impossibilities are best considered as singularities, only to be resolved if the dimensionality of the scene is first resolved. In an analogous way, depicting a vortex in three-space results in singularities (expressed as the need to divide by zero), which can be eliminated if the dimensionality is resolved.

After this break for a cup of coffee, let's return to the contemplation of the solar system as a vortex, a more useful task than trying to compute the behavior of a cup of coffee with cream and sugar.

The Dimensionality of a Fluid Vortex

Unless you really want to know more about the mathematical treatment of vortices, you can skip this section and move on to the next.

We have seen that a vortex is more than two-dimensional; indeed, this

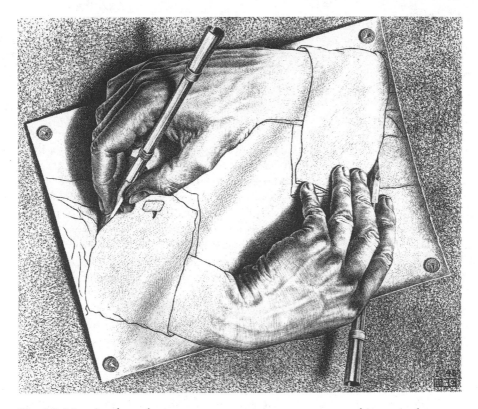

Fig. 5.2. Mapping from three-space onto two-space, as represented in a print by M. C. Escher. Reproduced from a print in the possession of Dr. T. R. Carlisle. (© 1994 M. C. Escher / Cordon Art-Baarn-Holland. All rights reserved.)

very fact is often included in the definition of a vortex. Put simply, if a fluid is spiraling inward on a surface (two dimensions), there is nowhere for the energy to go unless it leaves the plane of that surface.

Computations of the parameters of a vortex carried out in three dimensions, however, lead to a singularity of zero extent and infinite velocity at the vortex center—clearly an impossible situation. The center of a vortex is no black hole, though a black hole may be a vortex. Engineering practice provides empirical adjustments (fudge factors) to the calculations to avoid or "correct" such singularities. These singularities at the center of vortices are commonly ascribed to intermolecular forces and the stochastic turbulence resulting from these forces; some would call this quantum turbulence. It is difficult, however, to feel convinced that intermolecular forces and quantum effects can act across the 15 to 20 kilometers of the eye of a hurricane, the distances that would be required to suppress any calculated singularity and force the calculated eye of the hurricane to conform with observation.

Early in this century two mathematicians, the German Felix Hausdorff and the Russian Abram S. Besicovitch, independently used the nature of the singularity to prove that a vortex is less than three-dimensional. Their proof is related to the simple fact that the volume of a circle is zero. Put in another way, the extent of a circle, computed in three dimensions, is zero. A positive real value for the size of a circle is obtained only from calculation (or measurements) carried out in two dimensions; a circle has a finite area (two dimensions) but zero volume (three dimensions). To generalize, measurements or computations carried out in too high a number of dimensions will always yield an apparent size of zero and, in dynamic systems, an associated apparent infinite velocity. Finally, by extension, these mathematicians proved that any system of computation that yields zero size and infinite velocities has been carried out in too great a dimensional system. The converse is also true: computation in too low a dimensional framework will yield infinite extent and zero energy or velocity.

We have thus reached the impasse that a vortex is both more than two-dimensional and less than three-dimensional. Besicovitch emphasized the obvious implication that a fractional dimension, between two and three, was the only solution. (After all, there are plenty of numbers between two and three.) It was not, however, until the ready availability of computers and until Benoit Mandelbrot, a French mathematician, began to use them to explore the implications and properties of fractional dimensions (culminating in his book *Fractals* and in his discovery of the "Mandelbrot Set") that it became feasible to consider a vortex as anything other than a three-dimensional entity.

I have studied the appropriate dimensional framework for computing the parameters of a fluid vortex, and it turns out to be the logarithm of six divided by the logarithm of two, or approximately 2.58. Like π, this is an irrational number that continues indefinitely. Unlike π, no one is going to calculate its value to a million decimal places or more. On my pocket calculator I can come up with a value for this whirling dimension of the vortex of 2.5849625, a figure calculated to seven decimal places, and that is plenty for the purposes of calculation. An unending number like this is generally written with three dots at the end, indicating that it goes on forever: the whirling dimension of a vortex is thus written 2.58 . . . For our purposes we must regard a vortex as a 2.58 . . .-dimensional entity embedded in three-dimensional space.

Particle physics has been troubled with analogous problems of singularities. An electron, for example, is computed to be a point object—that is, an object of zero size carrying an infinite charge. Clearly, it is finite in extent (if finite has any meaning on quantum scales) and carries a charge of 1 electron volt; that is the definition of the electron volt. Many such singularities have been resolved by the use of what is known as renormalization group theory. The search for appropriate renormalization groups is tedious and empirical, and the resulting calculations are troublesome, but the final renormalization yields results that are experimentally testable. In principle, renormalization could be applied to the computation of vortices, but it is not obvious where to start searching for the appropriate renormalization group. Mandelbrot has, however, conjectured that fractal dimension and renormalization group theory are formally identical. In other words, they are the reverse of the same coin, being simply different ways of looking at the same phenomenon. It seems that singularities both of vortices and of particle physics could be resolvable by the use of fractional dimensions, if the correct dimensionality could be determined.

Dimension is an abstraction, one that has a variety of meanings not always compatible. Length, for instance, is a scalar quantity—that is, a quantity independent of direction. A piece of string is just as long however we lay it down on the table; its length is scalar. But if we distinguish length from breadth, then an element of direction enters into it, and length becomes a vector—that is, a quantity that possesses both size and direction. A scalar quantity can never be a vector, yet we use dimension in both senses. (Jargon rears its ugly head again, I fear.) We, as observers, may choose the form of abstraction that best suits our purposes. We may use length either as a vector or as a scalar quantity, but not as both at once.

For considering the properties of a vortex three distinct kinds of abstract dimensions, all vectors, must be distinguished: descriptive dimensions (also called degrees of freedom), projectional dimensions, and Euclidean dimensions. The first and last of these are integers; the second usually, but not necessarily, a fraction. Mandelbrot's formalism will not allow an integral value for this form of dimension, hence his term *fractal* for a subset of projectional dimensions. In the more generalized formalism I prefer, fractals are only a special case of projectional dimensions, and static geometries may have integral projectional dimensions usually identical with their Euclidean dimensions, so that the term *fractal* seems inappropriate. I present no formal definition of these terms.

The initial description of a vortex, as of any rotational movement, requires us to utilize six vectors, variously called degrees of freedom or dimensions. These I call descriptive dimensions, since they describe the motions of the particles that make up the fluid vortex. They are normally expressed as three Cartesian vectors and three rotational vectors, but they are symmetrical with one another, since a simple change of frame of reference will convert a rotational vector into a Cartesian vector, and vice versa. These six vectors are clearly distinct from the more usual Euclidean dimensions (or vectors), which are purely Cartesian in their expression. Such Euclidean vectors can never express rotation and in normal space are limited to three dimensions—generally, length, breadth, and height. The six vectors we require in a dynamic geometry to describe rotational movement hence require a different name—namely, descriptive dimensions.

It has often been proved that a fluid vortex (with six descriptive dimensions) can neither start nor end within the body of the fluid. It needs a spoon to start the swirl in the cup of coffee. The terminus of the vortex must always be on a solid surface (the inside surface of the coffee cup), which, like all surfaces, is two dimensional in the Euclidean sense. In a very real physical sense, then, a vortex represents (or may be expressed as) a mapping onto this surface, and it is a mapping from six (descriptive) dimensions onto two (Euclidean) dimensions—that is, a mapping from six-space onto two-space. The abstract projectional space in which the vortex is mapped is defined in terms of a third set of dimensions— projectional dimensions.

Mandelbrot has used proportional parts to derive the fractal dimension of self-similar curves such as the triadic Koch curve. He has shown that if a line segment is divided into r equal parts, each of which is then replaced by m identical portions (m is greater than r), with each portion being of the same form, then the fractal size of the resulting curve tends to infinity. The fractal dimension (or projectional dimension) of the resultant curve, whose construction started with a one-dimensional line on a two-dimensional plane, is the ratio of the logarithm of m to that of r. This can more easily be seen if we generalize the construction of a fractal line to the construction of a fractal plane. For a curve (or a plane figure) to have a fractal dimension in Mandelbrot's formalism, it is essential that it be constructed from self-similar parts whose similarity extends over a wide (or infinite) range of scales. The equivalent condition in renormalization group theory is the so-called scaling condition, which expresses much the same concept.

This condition is fulfilled in a vortex, which dissipates energy by cascading into ever smaller subvortices and eventually into turbulence. Moreover, in mapping from six dimensions onto two-space, the mapping must be valid at all scales. Thus I have found that Mandelbrot's principle of proportional parts applies to vortices, and the abstract projectional space, as I have shown above, has the whirling dimension of the logarithm of six divided by the logarithm of two ($m = 6$ and $r = 2$) or $2.58\ldots$

Computation of the parameters of vortices carried out with this value ($2.58\ldots$) rather than with three dimensions, yields no singularities, has no need for fudge factors (I am a scientist, not an engineer, and I prefer the scientists' terminology here as elsewhere, not that of the engineer, with its overtones of a cooking recipe). In addition, the computed values, even in crude calculations, compare well with observational scientific data.

The Earth and Moon as a Vortex

Recent theoretical work on the tilt of the axis of rotation of the earth and of the other planets, most recently by Jack Wisdom and his colleagues at the Massachusetts Institute of Technology and by J. Laskar and his associates at the Bureau des Longitudes in Paris, has suggested that the obliquity of the polar axis of planets is essentially chaotic. That is to say, the tilt of the axis of rotation varies in a chaotic manner over periods as short as a million years for the stony planets, with the axis of Mars in particular tilting as far over as 50 degrees at frequent (though irregular) intervals. Indeed, Mars may well have laid down on its side sometimes, with its axis of rotation tilted by 90 degrees so that it lies in the plane of the ecliptic at times. The earth is exceptional in that the moon stabilizes its tilt at approximately 23 degrees.

In my own studies I find that a consideration of the relationship of the earth and the moon, treating the system as a vortex, indicates that this satellite has indeed served to stabilize the earth's rotation, in particular the position of its axis. Throughout the duration of the partnership of the moon with the earth, the former has stabilized the axis of the latter to an angle near 23 degrees from the perpendicular to the plane of the ecliptic, or to put it another way, the equator is tilted by 23 degrees from that plane. To be sure, the poles of the earth are subject to nutation and precession, but this does not negate the stable angle of 23 degrees but merely allows it to wander around, causing the precession of the equinoxes.

Without the moon, the poles would be free to move, allowing the earth's axis to come to quite other angles. This has clearly happened to Uranus, whose axis of rotation at the moment lies just below the plane of the ecliptic, tilted at an angle of 97 degrees. Venus too must have passed through such a position, for it is upside down (actually at a tilt of 179 degrees) with respect to the other planets, with the south pole on top instead of the north, where it presumably started. Such retrograde rotation is more stable against chaotic influences than prograde (or direct) rotation, and the Venusian day has become linked in resonance with its year, thus stabilizing the current axis of rotation further. But even though Mars is currently rotating in much the same sense as the earth, it seems that its axis too has wandered all over the place, in the absence of any moon to stabilize it, and it exists currently in a completely chaotic state, though it has probably never turned completely upside down like Venus.

Consider for a moment what the climate would be like if the earth's axis lay in the plane of the ecliptic. During the northern summer the north pole would be pointing directly at the sun. Insolation would be equal to that of the present-day equator at noon, not just for an hour or two but for 24 hours a day. The pole would be the hottest place on earth, much hotter than any equatorial spot on the present earth. Six months later the south pole would be in the same position, receiving tropical sunlight 24 hours a day, and the north pole would be in constant darkness, much colder than it is today.

At the equinoxes, twice a year, the earth would receive sunlight much as it does at the equinox today. The equator would be in a summer condition twice a year. At the solstices the equator would receive very little sun, which would strike at a glancing angle, and would probably be as cold as the arctic is nowadays. Thus the weather at the equator would oscillate twice a year between what we today consider as tropical and arctic, while the polar temperatures would oscillate once a year between extremes far greater than any we experience at any point on the earth's surface today. The enormous temperature gradients would generate storms of unbelievable power, much stronger than class-five hurricanes. Could life survive these conditions? Could it even have originated under these conditions?

It is even possible that the runaway greenhouse effect that has overtaken Venus and led to the current high surface temperatures could have been occasioned by the period (or periods) in its past when its axis of rotation lay in the plane of the ecliptic. And perhaps the rills of Mars and the evidence of surface water at some time in its past could suggest that the surface had been subject to the much higher local temperatures created in a period when the axis of rotation was horizontal. All the evidences of surface water on Mars are polar in location, suggesting that the poles were once the hottest parts. All of this could be supportive evidence for the foregoing scenario.

To return to the earth, it seems clear that the presence of the moon is a necessary feature for stabilizing not just the angle of rotation but also the climate of the earth; hence the moon has had a great influence on the origin and evolution of life and of living things. The easy speculation that conditions for life must exist on countless planets around billions of other stars begins to seem less likely. Not merely is it necessary that a planet of the right size exist in the right location around a star, but also it seems that such a planet must have a large moon. It seems to me that this reduces immeasurably the odds for the existence of life as we know it elsewhere in the galaxy. Perhaps the Phoenicians were right in worshiping the moon goddess Astarte as the bringer of life and fertility!

The Solar System as a Vortex

The solar system may be regarded as a vortex, so long as we consider a vortex a collection of objects rather than a fluid continuum. Nearest the sun we have the four stony planets and the several million asteroids. Then comes the zone of the gas giants: Jupiter, Saturn, Uranus, and Neptune, together with Pluto, which orbits in this zone, and possibly the hypothetical tenth planet, planet X. Then comes the forbidden zone and finally the zone of the Oort Cloud, containing some trillions of comets.

Since three-dimensional geometry is capable of solving only the two-body problem and breaks down even with the "three-body problem," there is clearly no possibility of computing the behavior of all these objects in three dimensions. If, however, we treat this assemblage as a 2.58 . . . - dimensional vortex embedded in three-space, then the task, though laborious, is feasible.

Stars such as the sun continue to be formed today from giant molecular clouds when a supernova explosion starts a wave of compression in this gas. This is what is happening now in the nebula that bejewels the scabbard of Orion's sword in the constellation that goes by his name. The mass of gas in such a cloud is enough to lead to the formation of many thousands, even a million, stars, thus forming a star cluster, each starting from an eddy or vortex in the gas cloud stirred up by the explosion. All the stars in such a cluster are formed at roughly the same time (within a few million years) from the same starting materials and under the same conditions. A corollary is that if one such star has an Oort Cloud, then so must all the other stars in the cluster. The sun is one member of such a cluster of stars, and all its neighbors are members of the same cluster and of about the same age, all 4.5 to 4.65 billion years old.

There are two main theories of how the planets formed from the primordial nebula: the bottom-up theory and the top-down theory (charmingly descriptive names). The bottom-up theory posits that small "planetesimals" first condensed from the molecular cloud that gave birth to the sun and to the solar system. These planetesimals, in orbit around the sun, collided with one another, sometimes sticking (a process known as accretion) and sometimes blowing the accumulation apart, until the solar system as we know it was formed. This would be a comparatively slow process, taking millions of years. The top-down theory, which would lead to a much faster process, posits that great clumps of gas first condensed and that these were whittled away until we were left with the planets as they are now. For some unknown reason, the first condensations formed were too big to be stable and were broken apart, according to this theory, by internal tidal forces. In either theory the asteroids are formed from left-over bits unable to condense into a planet because of gravitational tidal forces exerted by Jupiter. Neither theory offers any hint of the origin of the Oort Cloud and of the comets.

All this neglects the role of differentiation of composition in the primordial star nebula. Clearly, the central sun condensed first, so that around it arose gradients of temperature and of gravity. The uncondensed part of the nebula, containing the greater portion of the rotation of the whole, rapidly differentiated in composition. The region later to become the zone of the stony planets received more than its fair share of oxygen, silicon, iron, and nickel and, for some unknown reason, an excess of carbon 12 (but not carbon 13). The region that was to become the zone of the gas giants is relatively lacking in these elements but has an excess of hydrogen and helium. The far outer reaches of the nebula retained the relative proportion of the elements initially found in the whole nebula, including a lower ratio of carbon 12 to carbon 13. At least, that is what I find when I try to develop a computer model of the early evolution of the solar system.

The implication of this zonal differentiation of the primordial star nebula is that the conditions for the formation of planets are quite different in these three zones. It could well be that the top-down model applies in one zone while the bottom-up model is more likely in another. Ernst J. Öpik, one of the principal proponents of the top-down theory, has shown that comets located in the Oort Cloud cannot be bigger than about 40 to 60 kilometers across. Let's see what happens if we recast his equations into the formalism of a 2.58-dimensional vortex. The results are portrayed in Figure 5.3 in the line marked "model 2." The figure graphically represents two computer models of the maximum size of accretion of the planets and other circumsolar objects. Model 1 is the bottom-up model of planetary formation; model 2, the top-down model. It can be seen that the top-down model (model 2) of the maximum stable size of objects at different distances from the sun fits the gas giants and the comets quite well but grossly overestimates the sizes of objects in the stony planet zone. The bottom-up model (model 1) fits the stony planets and asteroids.

The most recent implementation of the bottom-up model has been carried out by George Wetherill, who has used Monte Carlo simulation techniques to model the inner region of the solar system. His simulations, based (as are one-armed-bandit gambling machines) on specific rules

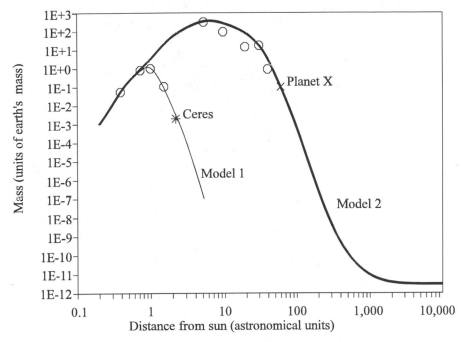

Fig. 5.3. Accretion of the solar system. Model 1 is the bottom-up model of planetary formation; model 2, the top-down model. Model 1 fits the inner stony planets
and asteroids quite well, provided Jupiter has formed earlier, while model 2 fits
the gas giants and comets. Ceres is the largest of the asteroids. The scientific notation used for expressing the size of the planets is explained in the glossary.

working within a framework of chance, have produced an excellent fit for
the four stony planets but have not provided any upper limits for their
size. Moreover, the model only works if Jupiter is already in existence,
which could not have happened if it too were formed by a slow bottom-
up mechanism. We must conclude that both mechanisms were in action:
the top-down condensation of big gobbets of gas for the gas giants and
the outer planets, and, in the rather different conditions of the inner stony
planet zone, the formation of planetesimals and their agglomeration into
stony planets and asteroids through the bottom-up mechanism.

If we use Monte Carlo techniques to compute a bottom-up vortex
model of the inner solar system in the presence of existing gas giants, using
the whirling vortex dimension of 2.58 . . . , then we obtain an upper limit
of size marked as "model 1" in Figure 5.3. Obviously, from the figure, this
is a far better fit for the stony planets and asteroids than is the top-down

curve. I agree with Wetherill that stochastic (chance) factors enter into regulating the precise position of planets and asteroids within the envelope formed by this upper limit of size.

What happened next? The planets had formed, and there were lots of bits left over—planetesimals, protoasteroids, protocomets. Can we come up with a calculation leading to a lower limit of size for objects in different parts of the solar system? Indeed we can, within the framework of a vortex, as long as we introduce chaos theory and carry out our calculations in 2.58 . . . dimensions.

It has been shown within the last few years that all the planetary orbits are in fact unstable and that the planets will ultimately either fall into the sun or wander off into outer space. It will take about ten billion years for this to happen—more than twice the age of the solar system—so it's nothing to worry about. Inevitably, these calculations have been carried out on individual planets, each one related to the sun as a two-body problem. By treating the solar system as a vortex we can deal with all the planets and smaller objects at once.

It turns out that the critical time for an orbit to go chaotic depends both on the distance from the sun and on the size of the orbiting objects. The smaller the orbiter, the faster its orbit decays into chaos. This is not a linear relationship but more of a stepwise function. Below a certain size, which depends both on the distance from the sun and the proximity to other planets, an orbit is stable for no more than about ten million years before it decays into chaos, while a larger object in the same orbit may be stable for ten billion years. The position of the step in this curve, as represented by the inflection point in Figure 5.4, makes a convenient marker dividing objects in stable orbits from those in chaotic orbits, marking the lower limit of size for an object to have persisted from the beginning of the solar system.

Assuming the present size and position of the planets (either including planet X, if it exists, or excluding it—it makes no significant difference), we can compute this lower limit for stable size. The result of this calculation is presented in Figure 5.5 as a line separating stable orbits (those lasting over ten billion years before decaying into chaos) from unstable orbits, which will decay into chaos within ten million years. Nothing below this line could have persisted to the present day. All planetesimals in the inner solar system left over from the formation of the planets have been swept clear, either by their bombardment of the planets (and the moon) in the so-called early bombardment episode or else by their wandering out of the solar system; very few would have hit the sun. In the region from approximately 80 to 2,000 astronomical units from the sun, the largest objects that could be formed are smaller than the stable range. This thus repre-

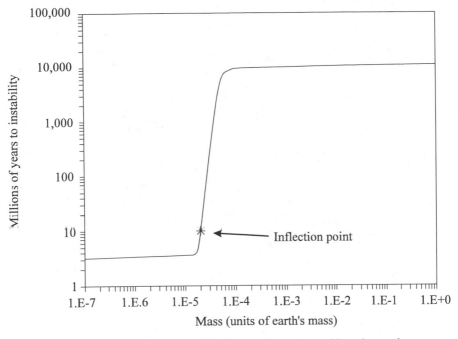

Fig. 5.4. Chaos in the solar system. This figure shows the sudden change from unstable, below about ten million years, to stable (where the period to the chaotic transition is more like ten billion years). The point marked "inflection point," where the curve changes from concave upward to convex, is a convenient marker and is used in subsequent figures to delineate the zones of stability and of chaos.

sents a forbidden zone where there are no stable orbits at all. The unstable Kuiper Belt (the source of short-period comets) is located in this forbidden zone.

The influence of Jupiter (and to a lesser extent of the other gas giants) not merely prevents the asteroids from condensing to form a planet but also determines that they shall have a lower limit of size of a little under a kilometer in diameter. This also accounts for the Kirkwood gaps, which can be seen in Figure 5.5 as the three spikes distal to the position of Ceres. The inner Oort Cloud starts at about 2,000 astronomical units, and the outer cloud at about 5,200, being separated by a gap analogous to the Kirkwood gaps from the inner cloud. Similarly, the lower limit of size for objects in the Oort Cloud is about 100 meters. Below this diameter an object will coalesce with other objects or else be lost to the solar system.

Now at last we can see why the zone outside the planetary zone but nearer to the sun than the Oort Cloud may be called the forbidden zone:

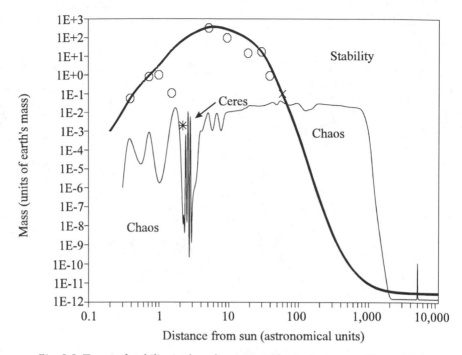

Fig. 5.5. Zones of stability in the solar system. This is the same as Figure 5.3 (omitting model 1), with the addition of the line separating the chaotic zone from the stable zone, derived from the locus of the inflection point shown in Figure 5.4.

the largest objects that can form here are below the critical size for stable orbits, so nothing can remain here for very long. (Time is relative: "very long" here means more than a few million years.) I suggest, though I have no proof, that the forbidden zone is the source of the objects responsible for the second era of bombardment of Mars and the moon (called the "terminal lunar cataclysm"), which ended 3.8 billion years ago. Objects formed in this zone were in orbits only stable for a few million years, and many would have eventually wandered into the inner solar system.

By this astronomical era, then, the solar system looked very much like it does today, with an inner zone (the planetary zone, devoid of small orbiting objects except for the asteroid belt), a forbidden zone, and then the zone of the Oort Cloud. In addition, all other stars in the cluster resulting from the same star-forming event (perhaps as many as a few thousand) were in the same state, each with its own Oort Cloud and its planetary zone, whether or not actual planets had formed.

Let's look at the Oort Cloud in rather more detail. Figure 5.6 shows the

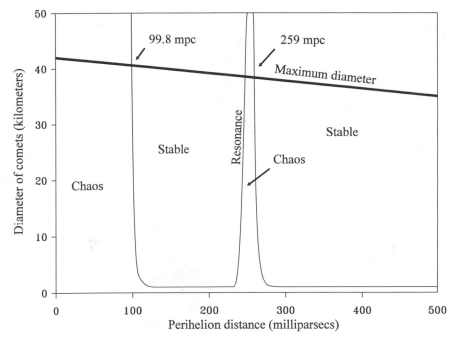

Fig. 5.6. Zones of stability in the Oort Cloud. This figure shows the Oort Cloud in an analogous form to Figure 5.5 of the whole solar system. The axes are linear instead of logarithmic, and the whole planetary system, which lies in the first couple of milliparsecs, is omitted. Size is expressed in diameter of comets instead of mass, and distances in milliparsecs instead of astronomical units.

same kind of calculations (2.58 . . .-dimensional) I have used in the preceding figures, with a maximum size dictated by dynamic considerations and a minimum size by considerations of instability of orbit. The inner Oort Cloud has a well-defined inner limit at 99.8 milliparsecs from the sun, and a not-well-defined outer limit (separated from the inner by a forbidden zone) at 259 milliparsecs. This limit depends on the nearness of adjacent stars and varies with time as stars come and go. These are average figures, since the Oort Cloud is distorted by the gravitational pull of the center of the galaxy. The nearest star (Alpha Centauri) lies about 1.2 parsecs (1,200 milliparsecs) from the sun at the present, and the outer Oort Cloud extends halfway to this star. There is a clearly defined gap, comparable to the Kirkwood gaps of the asteroid belt, dividing the Oort Cloud into inner and outer Oort regions. The maximum diameter of comets is about 40 kilometers, in agreement with Öpik's value, corresponding to a mass of about 10^{19} grams. The minimum diameter of a comet for a stable orbit is about 100 meters. The orbital period—the cometary year—is about a million earth years. It is from the Oort Cloud that the comet (or comets) came to disrupt life on the earth at the Cretaceous-Tertiary Boundary.

Asteroids and Comets

We have obtained some indications of the composition of both asteroids and comets by telescopic means, chiefly by the use of the spectroscope. Far more, however, is known from the study of meteorites, most of which represent fragments of asteroids. This is no place for a detailed discussion of the chemistry of these objects. A few salient facts are necessary, however, to the main argument of this book. (I will not bore you with the details, though to me they are fascinating—but then I'm not normal.)

The asteroids were formed in the zone of the stony planets—that innermost region of the primordial star nebula that is enriched in oxygen, silicon, iron, and nickel—and, *taken as a whole*, have the composition of the earth. Individually they vary enormously in composition. Agglomeration and coalescence have led to collisional and gravitational heating so that the separate minerals of which they are composed have been able to segregate, the heavier iron and nickel forming cores, with a slag of siliceous minerals on the outside. Subsequent collisions have broken them all apart and readhered them repeatedly. Asteroids can be all stony (the slag from the surface of the original planetoids) or all metallic (the original cores) or they can be any mix in between. Accordingly, meteorites too may be rocky or metallic and show all possible combinations of the primordial material.

Asteroids (and the resulting meteorites) originated in the zone of the stony planets, that region enriched in carbon 12. Accordingly, they exhibit the carbon-12-to-carbon-13 isotope ratio characteristic of terrestrial carbon, roughly 90:1. This is not true of the carbonaceous meteorites, which, though their trajectories suggest they came to earth from the asteroid belt, did not originate there. I will show later that these originated in the Oort Cloud and have been parked temporarily in the asteroid belt (or in the Kuiper Belt) for no more than a few million years. They have a carbon isotope ratio closer to that of interstellar dust (at least dust in nearby interstellar space), with a carbon-12-to-carbon-13 ratio of about 42:1. In other words, they have less than half as much carbon 12 as earth and the asteroids. Even though they are present in the asteroid belt, they are not original asteroids.

In contrast to asteroids, comets were formed in deep space, around the edge of the primordial star nebula, and have never been heated by the sun. They have been likened to "dirty snowballs," consisting chiefly of ices (water ice, methane ice, and others) with 10 to 20 percent of "dirt," mostly carbon (both graphite and diamonds), organic compounds, and a minor mineral component. Since the region where they were formed was deficient in oxygen, they contain compounds impossible to form in the asteroid belt and lack others common in the inner solar system. Silicon (the element) has not been oxidized to silica (silicon dioxide, or quartz) and to the siliceous rock minerals common on earth and the asteroids. Instead, most of the silicon is in the form of silicon carbide, which on earth exists only as a man-made material. Other minerals rare on earth but common in comets include spinels and chromite.

Comets today are found chiefly in the Oort Cloud, whose inner edge is about 21,000 astronomical units from the sun. They move relatively slowly in orbit and take about a million years to complete a single turn; in other words, a comet's year is about a million earth years. Only if they are perturbed in some way will they swing in toward the sun and become visible as "hairy stars"—and present the threat of colliding with the earth.

As a comet swings in toward the sun it is heated at the surface and the ices sublimed off to form the "hair," the tail of the comet. Accordingly, the "dirt" remains behind to form a black crust on the surface. From direct telescopic observations of comets we know far more about the tail, and hence about the volatile substances, than about the solids left behind by sublimation. A summary of the chemical differences between asteroids and comets follows:

	Asteroid	Comet
Carbon 12 to carbon 13 ratio	89 : 1	42 : 1
Silicon	Oxides	Carbides
Heat differentiated	Yes	No
Free carbon	No	Diamonds, graphite
Organic compounds	No	Amino acids, aromatics
Spinels	Rare	Yes
Chromite	Rare	Common

The Formation of Asteroids from "Dead" Comets

Many forms of perturbation of the Oort Cloud can lead to the entry of comets into the inner solar system, though few will actually penetrate through to the innermost zone of the stony planets. Unless they are moving exceptionally fast, most comets will reach a parking orbit either just outside Neptune, in the Kuiper Belt, or in the asteroid belt, being shepherded there by gravitational effects of Jupiter and the other gas giants. In the belt outside Neptune they remain unheated by the sun and so retain all their ices and volatiles, but they are in unstable orbits. In this region there is no such thing as a stable orbit; all orbits are chaotic. In other words, comets can only remain there for a few million years, characteristically about ten million, before the orbit degenerates into chaos and they are expelled. At this point many of them will start to wander out again, often to be lost to the solar system, but others will move in quasi-elliptical orbits into the inner solar system. These are the short-period comets, comets with a return period of less than 200 years, such as comet Halley (period: 76 years). As they swing in toward the sun, sublimation of the ices begins and tails develop.

Comets shepherded into the asteroid belt by the gravity of Jupiter undergo a different fate. They have entered the region where solar heating will expel all the volatile substances; what is left is a dead crust of carbonaceous material. This is the origin of the carbonaceous asteroids: they did not originate in the asteroid belt but in the Oort Cloud. This too is the origin of the carbonaceous meteorites that have come to earth along trajectories originating in the asteroid belt. They are fragments of comets that have lost their volatile fraction and have been parked in orbits in the asteroid belt for a period, where they have been baked to a crisp.

The Graveyard for Dead Comets

What happens to comets that enter the inner solar system from the Oort Cloud? There are several possibilities.

First, they could just be expelled from the solar system on quasi-hyperbolic orbits. This sounds plausible, but such orbits are extremely difficult to observe and are almost impossible to distinguish from extremely elliptical orbits. As a result, no comet has ever been observed moving on an unequivocally hyperbolic trajectory. In addition, it is difficult to model any plausible event that might result in such a trajectory.

Second, a comet could hit something, such as the earth or the moon. It could not hit Jupiter, whose escape velocity is too high, and it could not fall into the sun for the same reason, despite occasional observations of sun-terminating comets. The most likely explanation for such observations is that they represent sun-grazing comets which make very close approaches to the sun, thereby evaporating all the volatiles, which form the tail and normally make a comet visible. The result is that a small comet would no longer be visible without its tail as it speeds away from the sun.

Third, a comet could be parked in an orbit beyond Neptune. A resonance with the gas giants creates a trap there, about 100 astronomical units out, where comets may be confined for a few million years, the so-called Kuiper Belt. Indeed, my computer models suggest that this is the fate of the majority of comets perturbed inward from the Oort Cloud. This belt, however, is unstable; after about ten million years the trajectory of anything orbiting there decays into chaos, generally into a quasi-elliptical orbit that brings it into the zone of the stony planets. Such a comet is known as a short-period comet—that is, one with a return period of less than 200 years—and is designated with the prefix P (for periodic). P/Halley and P/Swift-Tuttle are examples of comets whose current orbits originated in this way, after being parked for a few million years in the Kuiper Belt. This belt, then, is not actually a graveyard for comets, but rather a long-term parking lot.

Fourth, comets may be trapped by resonance with Jupiter in the asteroid belt. This seems to be the ultimate fate of the majority of comets. Here they are within the range of the sun's heating rays, which will boil off all the volatile materials comprising the ices that form the characteristic tails of comets. A comet that has lost all its volatiles and no longer has a tail is said to be a dead comet. The asteroid belt is truly the graveyard for dead comets.

A dead comet consists of only the nonvolatile remnants of the original material, 80 percent of which has been lost, boiled off by the heat of the sun, leaving only a dehydrated skeleton. Chondrules of interstellar material—inorganic and organic carbonaceous materials, including amino acids and primordial dust (including interstellar diamonds)—are all that remain. The dead comets have become asteroids, not primordial asteroids formed in situ, but bodies trapped there later, often much later, by Jupiter's gravitational field.

Accordingly, today when we use the telescope to measure the spectral properties of asteroids, we can distinguish several different kinds. Those whose spectrum corresponds well with stony meteorites are either whole undamaged asteroids, with a metallic core and a stony mantle, or else fragments of asteroidal mantle. We may suspect that none of the former actually exist, since the tidal disruption caused by Jupiter's gravity has long ago fragmented any primordial asteroids. Those whose albedo and other properties agree well with iron/nickel meteorites are fragments of asteroidal core material. And those whose spectral properties are those of carbonaceous meteorites are fragments of dead comets.

Most of the meteorites that hit the earth come from the asteroid belt, fragments perturbed from their stable orbits by collisions hard enough to break them apart, collisions that occur repeatedly. The fragile nature of the material of dead comets means that collisions are more likely to lead to fragmentation than is the case for the tougher rocklike or metallic "native" asteroids, those asteroids born in the asteroid belt.

All this means that although carbonaceous meteorites hitting earth have come most recently from the asteroid belt, their ultimate origin was from comets from the Oort Cloud. Thus, carbonaceous meteorites are simultaneously asteroidal in origin (in the sense that they come to earth from the asteroid belt) and fragments of dead comets. Their very composition shows this. Primordial diamonds, of supernova origin, could not have survived the melting and differentiation that native asteroids have undergone, nor could the organic materials with which they are loaded. Their ages, as determined by radioactive dating techniques, are older than those of any other meteorites and have been taken as an indication of the age of the solar system—some 4.54 to 4.56 billion years. No stony or metallic meteorite is that old, in the sense that their radioactive clocks have been reset by the heating they have undergone in the differentiation of the parent body into core and mantle.

The chief graveyard for dead comets, then, is the asteroid belt, where their bones do not rest in peace but are kicked around by the asteroids and by tidal forces from Jupiter's gravitation until they fragment. The resulting fragments—the carbonaceous meteorites—are our best guide to the composition of comets and to the composition of the nonvolatile fraction of the primordial nebula from which the solar system condensed. They are the oldest objects to which we have had access.

Chaos

Chaos—that state of things determined by chance alone—is a potent force, one that has shaped the solar system as we know it today and that will ultimately, in another five billion years, lead to its disintegration. Chaos has defined the "forbidden zone" between Neptune and the Oort Cloud, including the Kuiper Belt. Chaos throws short-period comets out of their unstable orbits beyond Neptune into earth-crossing trajectories. Chaos is at the heart of the dynamics of the solar system, as noted in Figure 5.5. As the American composer John Cage wrote in 1961 in his book of essays, *Silence*, "Let us say *Yes* to our presence together in Chaos." Although to John Cage chaos took on Zen Buddhist overtones of meaning, which is different from a mathematician's more rigorous usage, nonetheless the message is the same.

Chaos rules our weather, our solar system, and our very lives. Since we cannot escape chaos, let us embrace it: "Let us say *Yes* to our presence together in Chaos." Even if we cannot actually embrace chaos in our daily lives, it is the chaotic behavior of comets that has led to mass extinctions in the past, to our very existence today, and perhaps to our own eventual extinction when the next comet hits the earth.

Bolide Impacts and Vulcanism

Incensed with indignation Satan stood
Unterrified, and like a comet burned
That fires the length of Ophiucus huge
In the arctic sky, and from his horrid hair
Shakes pestilence and war.
—John Milton, *Paradise Lost*

Man has always feared the apparition
of comets, even into the present century. The very word *apparition*, the
term for the appearance of a visible comet, expresses some of this dread,
having overtones of the supernatural. The fear of the earth's passing
through the "hair" of the comet during the 1910 apparition of comet Halley provoked a panic in London and New York. Comet Halley also appears as a portentous object in the Bayeux tapestry as foretelling the Norman Conquest of England in 1066; a millennium earlier the comet was
mentioned with fear and loathing by Chinese astronomers. The 1456 apparition provoked panic in Europe because it was thought to portend the
Muslim invasion after the fall of Constantinople in 1453, while in that
city (which was by then called Istanbul), the Ottoman Turks feared it portended the city's recapture by the Christians.

The comets of 1664 and 1665 were held in Britain to presage pestilence
and fire, and indeed, an epidemic of plague in 1665 and 1666 was followed by the Great Fire of London in which half the city burned. Even
Isaac Newton fled London, but whether this was because of the omens (he
was after all a great believer in omens and numerology) or because of the
plague itself is uncertain.

The apparition of comet Halley in 1682 during the fateful reign of King
Mbakama-Mbom in central Africa and the tribal warfare that followed
are still remembered with fear in the oral tradition of the Kuba Confedera-

tion of Congo/Zaire. The 1835 apparition of comet Halley over central Africa presaged to many Africans the cataclysmic events that were to follow—epidemics of measles and smallpox, warfare, famine, and slavery—in the reign of King Mbopy Mabiintsh Ma-Mbul, when European penetration of this area really began in force.

For a period of two months in 1882 all newborn children on the coast of Tanzania at Uzigua were sacrificed to placate the gods when a particularly brilliant comet (the Great September Comet, 1882 II) was visible even during the day. This apparition is still remembered with fear from Tanzania to Nigeria, while the sacrifice of the children cannot help but recall to the Western mind the appearance of a star in the East and the Herodian Slaughter of the Innocents.

But it is not the hair of the comet we must fear but rather its head. The recent rediscovery of P/Swift-Tuttle (named for Swift and Tuttle, the American astronomers who first described it) has led to speculation that there is a real risk of its hitting the earth on August 14, 2126. Indeed, some astronomers have calculated the likelihood of this happening during the three minutes in which the comet is actually in the orbit of the earth as one in 10,000. This probability figure, though, makes some assumptions that are totally unacceptable.

P/Swift-Tuttle is a dirty comet, constantly spewing off particles of dust and debris when it is near the sun every century and a half. As a result, its former track is marked by a veritable river of dust. This lies across the orbit of the earth, and every August 11 (getting slightly later by a few minutes each year) the earth actually passes through this river. Particles of dust enter the atmosphere and are seen as shooting stars or meteors as they burn up. The critical assumptions involved in the probability calculations are that (1) P/Swift-Tuttle will follow exactly the same track as at its last apparition, when it shed the dust we see as meteors; (2) the earth's orbit has not been perturbed by the smallest fraction since the last apparition; and (3) the orbit of the comet will not bring it anywhere near the vicinity of any of the gas giants (Jupiter, Saturn, Uranus, or Neptune) either in the current apparition or the next, in 2126.

If we grant these three assumptions, then the task of calculating the probability of the comet hitting the earth becomes a simple one that can be treated as a problem in "probability cross section." This is a field of mathematics that first came into prominence in constructing the atom bomb. Nowadays solving a problem like this can be carried out in a few minutes on a desktop computer. I calculate it (no doubt to a spurious degree of accuracy) as one chance in 13,149. But these assumptions are absurd. First, every comet passes within gravitational range of one or more gas giants and is perturbed by this close approach. Thus it is impossible to predict the time or even the year of return of a comet. There may be as much as a three-year error.

Second, when it is warmed by the sun a comet sends out jets of gas and vapor that act as powerful jet engines to propel it erratically out of its precise former trajectory. Third, we know that the earth's orbit has undergone numerous minor perturbations over the years. The eruption of El Chichón volcano, for instance, caused a measurable alteration in the orbit of the earth—measurable perhaps only in the millisecond that must be added to the theoretical length of the year, but enough to upset the sort of calculations required by doom predictors. Each time a volcano like Pinatubo erupts, the odds of P/Swift-Tuttle hitting the earth change, usually becoming longer.

This sort of chaotic behavior is impossible to calculate, but a back-of-envelope estimate suggests that the odds of P/Swift-Tuttle actually hitting the earth in 2126 are more like one in many millions rather than one in ten thousand; the precise date of closest passage is impossible to predict this far ahead, and even the year is in doubt.

Bolide Impacts

What happens when an object from outer space (the general term is *bolide*) hits the earth? A small bolide is slowed down by its passage through the atmosphere and either burns up completely in the atmosphere as a meteor or else lands on earth, where it makes no crater. A larger bolide, with a lower surface-area-to-mass ratio, is not slowed down as much and will form a crater when it hits.

One of the best-known examples of a crater formed in this way is the Barringer Crater (also known as the Meteor Crater), which is a bowl-shaped depression in the desert of northern Arizona (Figure 6.1). The bowl is about 1,200 meters across and 170 meters deep, with an elevated rim. It was formed 30,000 to 40,000 years ago by the impact of an iron meteorite about 40 to 60 meters across and weighing about a million metric tons. The bolide hit the earth at a relative speed of about 15 kilometers per second and released some 10^{17} joules of energy, roughly equivalent to a 20 megaton hydrogen bomb. Below the floor of the crater is a "breccia lens" consisting of broken rocks that have been changed and stuck together by the high temperatures and pressures of the impact—the process of shock metamorphism. This breccia lens in the Barringer Crater is about 380 meters thick.

When this bolide hit the Arizona desert floor the kinetic energy of motion produced a pair of shock waves, one through the bolide itself, heating

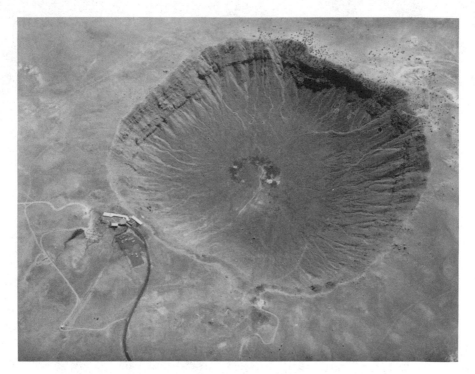

Fig. 6.1. The Barringer Crater, or Meteor Crater, Arizona, an example of a simple impact crater. Reprinted by permission, R. A. F. Grieve, the Geological Survey of Canada, Department of Natural Resources.

the body and finally blowing it apart, and the other down into the desert floor. This wave was followed by the so-called release wave, which decompressed the target rocks that were compressed by the first wave. Rock below the point of impact was fractured; the release wave traveled faster than the compression wave and overtook it, creating complicated dynamics. Broken rock was pushed downward and outward by the two waves, whose interference pattern forced some of this moving material upward and outward. Some material was thus ejected from the center of impact, and other rock, fractured by the impact but not ejected, was pushed up to form the elevated walls. The downward movement of the rock compressed by the first wave, combined with the ejection of material from the center, creating a transient cavity lined with fractured rock, which then collapsed inward, forming the breccia lens. The curtain of ejecta thrown into the atmosphere by the impact rained down all around, covering the floor of the crater (the breccia lens) and the surrounding desert with a blanket of ejecta over a diameter of many kilometers. Only the small fraction of ma-

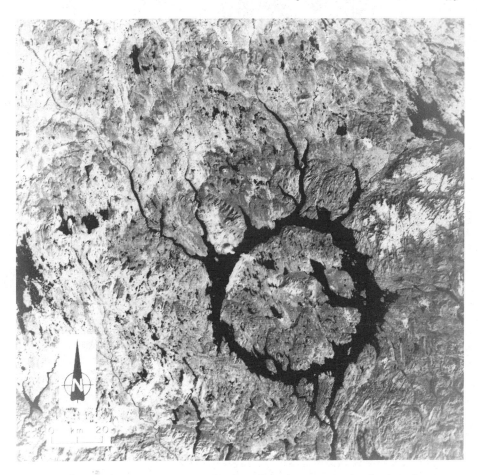

Fig. 6.2. The Manicouagan structure, Quebec, Canada, an example of a complex impact crater. Reprinted by permission, R. A. F. Grieve, the Geological Survey of Canada, Department of Natural Resources.

terial actually vaporized (including most of that of the bolide itself) traveled any great distance.

Simple craters like this may be as large as 4 kilometers across in igneous rocks, perhaps half that size in sedimentary rocks. Any crater larger than that is much more complex in form. A "small" crater, such as the Barringer Crater, may be about seven times as wide as it is deep. A larger crater, such as the 210-million-year-old Manicouagan in Quebec (Figure 6.2), initially 100 kilometers across but now reduced to 75 kilometers, may be 100 times wider than deep as a result of gravitational rebound in the center. Craters like this have an uplifted central structure—like the central cones seen on many large lunar craters—with a deeper circular trough inside the rim; the Manicouagan Crater at the present day is an annular lake.

Manicouagan was probably formed when a bolide, about a kilometer across and traveling at about 30 kilometers per second relative to the earth, hit the Precambrian rocks of the Laurentian Shield. The energy released was about 10^{21} joules, about 10,000 times that released in the Barringer impact. The compression wave did not merely fracture the bedrock but melted a portion. The release wave created a transient cavity, but under that the rocks that had been compressed by the impact rebounded upward in the center and the surrounding brecciated rocks flowed inward. The much more extensive blanket of ejecta contained molten droplets, formed as the rock melted and known as tektites (Figure 6.3). The rim of the cavity slumped down in a succession of bites to add to the brecciated lens on the floor of the crater. A complex crater of this kind does not have a raised rim, like the Barringer Crater, but instead has a raised central feature or cone.

Most impact craters have eroded over the years, becoming vaguely circular features. The form alone is often not enough to guarantee that any particular feature of the landscape marks it as an impact crater, though modern seismic methods, combined with drilling, have produced better ways of determining this. The Manson structure in Iowa, for instance, is totally invisible at the surface, and only modern methods have revealed its presence, showing that it is indeed an impact crater about 35 kilometers across, with an estimated age of 65 million years. Only the youngest craters have retained anything like their original form.

Diagnostic Features of Impact Craters

There are two diagnostic features of impact craters that have never been seen in any other geological formation. Brecciated rocks, such as the breccia lens of an impact crater, are not unique to these craters, and a volcanic caldera (or basin) may have the same shape as an impact crater. Shock-metamorphic effects and tektites, however, are the only geological features known to have been produced solely by impact cratering. The pressures produced in even the most violent volcanic event are simply too low to produce shock metamorphism. Even the pressures during orogeny, or mountain-building, never exceed one gigapascal. The peak pressure in as small a crater as the Barringer may have reached 10 gigapascals for a few score milliseconds, while in Manicouagan the peak pressure may have been as high as 80 gigapascals.

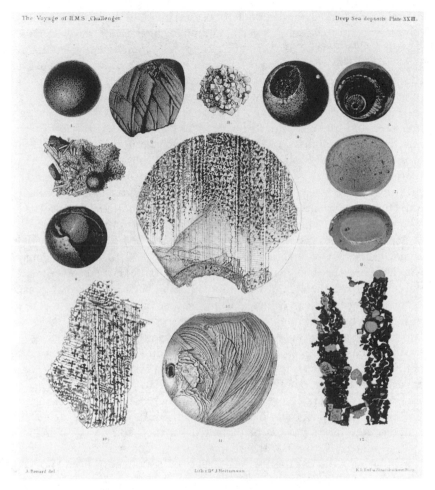

Fig. 6.3. Tektites and small meteorites from abyssal depths of the oceans. These are the earliest such finds published. Reprinted from the *Challenger Reports, Deep Sea Deposits* (1887).

To the naked eye the most obvious shock effect is the formation of shatter cones, which form at pressures between 5 and 10 gigapascals. These cone-shaped rocks, fractured in striated patterns, are best known from the brecciated lenses below terrestrial impact craters. Their presence was expected on the moon, but they have not yet been observed there.

Other shock effects are seen only under the microscope, but these have been reproduced in the laboratory. Grains of quartz and of feldspar, so long as they are not melted to form tektites, develop finely spaced arrays of fractures oriented parallel to major crystal faces. In the laboratory these have formed at pressures between 7.5 and 25 gigapascals, never at lower pressures. The density of the fracture planes and their orientations are directly related to the pressure of formation, so that naturally occurring shocked quartz grains can be compared with laboratory standards to estimate the pressure at which they were formed.

Diamond can also be produced from other forms of carbon under high pressure, and the mineral stishovite can similarly be produced from quartz. Volcanic action can, of course, produce diamond, and that is how most diamonds are formed, to be found later in the kimberlite pipes that represent the deeper parts of the central lava pipes of a particular kind of extinct volcano. Most diamonds of commerce are at least two billion years old. Diamonds found in the breccia lens of impact craters have probably formed in situ, but their presence is hardly diagnostic of impact.

More diagnostic is stishovite, a highly unstable form of quartz that could be formed within the earth at depths of 500 to 600 kilometers, where the pressure may reach 16 gigapascals. It could not, however, reach the surface from these depths without changing back into normal quartz as the pressure is released slowly. Rapid release of pressure over a course of less than a second could stabilize stishovite, and that is just what we find in the breccia lens of an impact crater and in the blanket of ejecta.

Above about 30 to 40 gigapascals feldspar and quartz lose their crystalline structure entirely and are converted into glassy replicas of the original crystals. These too have been found in association with known impact craters, but nowhere else. Above about 60 gigapascals most rocks melt. The melt from impact is not like the lava of a volcano. First, the melt often contains copious fragments of unmelted bedrock composed of higher-melting materials. Second, the chemical composition is quite different, consisting of a mix of the various target rocks that have been shocked above their melting point.

Perhaps the most commercially viable impact crater is the Sudbury Basin in northern Ontario, Canada—the world's richest nickel deposit. The basin represents the remains of an impact of a bolide 150 to 200 kilometers across that hit about 1.82 billion years ago at a relatively low velocity and must have a caused a major extinction in the Proterozoic era. The pool of melted rock was so huge that it did not solidify for about a million years and had time to differentiate. Around the cooler outer edge heavy nickel sulphide minerals, probably originating from the bolide, crystallized out first, making what is today a saucer-shaped ore deposit that is the richest nickel deposit in the world. A crust or slag formed on top of the pool, consisting of light minerals such as quartz. The interior of the pool, containing minerals of intermediate density, crystallized last, forming an overburden above the nickel deposits. The bolide that produced the Sudbury structure was probably a fragment of the metallic core of an asteroid, containing much nickel, with some iron, platinum, iridium, and cobalt. These so-called siderophilic elements are not common in crustal and mantle rocks of the earth but are concentrated in the earth's core, as they also appear to be in the core of asteroids. The presence of siderophilic elements has often been taken as a good indicator of extraterrestrial impact.

In the final analysis the best diagnostic characteristic of an impact event lies in the ratio of isotopes of certain elements. The relative abundance of isotopes of some elements—notably carbon, osmium, xenon, and rhenium—differs in the earth's crust from what is found in extraterrestrial environments. Isotope analysis can thus offer important clues to the origin of a crater and has been important in the researches described in this book.

The Cretaceous-Tertiary Boundary Event as an Impact

Luis and Walter Alvarez and their colleagues at the University of California at Berkeley suggested in 1980 that the Cretaceous-Tertiary Boundary event was caused by the impact of an asteroid bigger than the one that produced Manicouagan but smaller than the Sudbury asteroid. They pointed out that the discontinuity in the fossil record at the Cretaceous-Tertiary Boundary seemed more abrupt when paleontologists looked at the fossils of smaller creatures, which produce more abundant fossils than do larger ones. But just how abrupt was the extinction event? Peter Ward and his colleagues at the University of Washington first studied the extinction record of ammonites at Zumaya on the north coast of Spain and concluded that it took about five million years between the first noted decline and the final extinction. Then they looked at rocks exposed on the beach near Biarritz in France. (It was a nudist beach and the geologists were usually the only people wearing clothes; they confess that they were often distracted.) There, by painstaking stratigraphical studies, they pinpointed the extinction event to a period lasting no more than 100,000 years and more probably to as little as 10,000 years, which is a span of time longer than the entire course of recorded human civilization. Ward believed that the longer period he had estimated in the Spanish rocks was an artifact of a locally incomplete fossil record.

One of the classical sites for the study of the Cretaceous-Tertiary Boundary is at Gubbio in Italy, first discovered by Isabella Premoli Silva of the University of Milan. Alvarez and his coworkers studied the Cretaceous-Tertiary Boundary claystone within this marine limestone deposit and concluded that it represented an interval of no more than 1,000 years. Finally, Jan Smit of Amsterdam carried out a study of a rock section at Caravaca in southern Spain and concluded that the boundary claystone was deposited within a period of at most 50 years.

The Gubbio rock section figured prominently in the early studies of the Cretaceous-Tertiary Boundary. Alvarez and his colleagues found a higher level of the siderophilic element iridium in the claystone than in the rocks above and below—over 100 times the concentration. Iridium is rare in the earth's crust, and even in volcanic products. After testing several alternative hypotheses they finally suggested that an asteroid about 10 kilometers in diameter and containing about 10^{18} grams of material had hit the earth and dumped an enormous amount of iridium and other siderophilic elements into the atmosphere.

Since then scientists at a score of laboratories in a dozen countries have reported anomalously high levels of iridium at the Cretaceous-Tertiary Boundary at nearly a hundred sites throughout the world, in terrestrial deposits as well as marine and in ocean sediment cores. Gold, platinum, osmium, ruthenium, and rhodium anomalies have all been found at the same horizon; these too are siderophilic elements that are rare in the Earth's crust but relatively abundant in meteorites.

In fact, in studies at the Stevns Klint section in Denmark, at the section from Turkmenia, and from elsewhere, these platinum-group elements are found in the Cretaceous-Tertiary Boundary clay in the same ratio as they occur in a certain type of meteorite, the carbonaceous chondrites. Spread out over the world, the total amount of them all corresponds to the amount that would be found in a metallic asteroid fragment of 10^{18} grams.

Further evidence comes from isotope ratios of these elements. The rare element osmium is found in the earth's crust in a mix of two main isotopes, osmium 186 and osmium 187. The latter is relatively rare in meteorites compared with the ratios in terrestrial osmium; meteorites are enriched with the former isotope, and so it is in the Cretaceous-Tertiary Boundary clay.

Shock metamorphic features are also abundant in the Cretaceous-Tertiary Boundary. Jan Smit discovered spherules of altered basalt in the boundary clay at Caravaca; they have since been found in enormous abundance at several sites in the Caribbean and in the Yucatán peninsula of Mexico. These spherules are the basaltic equivalent of tektites, which are primarily formed from silicate rocks, and their chemistry suggests that the impact in which they originated took place on oceanic crust—that is, under water, not on land. "Under water" is a relative description. A bolide

10 kilometers in diameter would stick out of the ocean by more than half its diameter when it touched the bottom, even if it landed in the abyss. The evidence suggests that these spherules originally were glassy in nature, formed from melted basalt, but have largely turned to clay over the years. Some of them still have glassy cores, and those that have been dated were formed at a period of time indistinguishable from the age of the Cretaceous-Tertiary Boundary.

Shocked grains of quartz have also been discovered in a number of rock sections, including the Red Deer Valley section in Canada. From the spacing of the intersecting bands of lamellae, we can estimate that these were shocked at pressures around 25 gigapascals.

Finally, the clay itself of the Cretaceous-Tertiary Boundary layer is unlike most clays. The mineral illite, the most common constituent of normal clays, is replaced by smectite, and the total composition of the Cretaceous-Tertiary Boundary clay corresponds to a mix of twelve parts of altered basaltic material to one part of meteoritic material.

Volcanic Activity at the Cretaceous-Tertiary Boundary

I will reveal my bias immediately: I don't believe for one moment that enhanced volcanic activity is the cause of the Cretaceous-Tertiary Boundary event. I have made no effort to hide my bias. If you want a more positive account of why some people believe that vulcanism is the cause of the Cretaceous-Tertiary Boundary events, read their own accounts. We do know that the end of the Mesozoic was a period of high vulcanicity, especially in India. The Deccan Traps of southern India, now eroded away to less than half a million cubic kilometers, represent the remains of massive volcanic flows that took place over a period of a million years around 65 million years ago, the time of the Cretaceous-Tertiary Boundary event. (In the usual way that we mix languages and systems of terminology, *Deccan* means "southern" in Sanskrit, while *trap* comes from a Dutch word for staircase; the area was given the name we use in English by the Dutch East India Company.)

The sedimentary rocks immediately below the Deccan Traps contain fossils that are known only from the last million years of the Cretaceous era, including fragments of dinosaur fossils; these are absent from sediments above the traps. The vulcanism obviously spans the Cretaceous-Tertiary Boundary, so the lava flows are of the right age and are certainly associated with the Cretaceous-Tertiary Boundary in some way.

I find the arguments of those who propose that the Deccan Traps eruption was the cause of the Cretaceous-Tertiary extinctions quite specious, almost special pleading. (I did say I am biased didn't I?) Some of the spherules of altered basalt that are quite common in the Cretaceous-Tertiary Boundary claystone *could* be artifacts, such as insect eggs or modern algae. Shocked quartz grains *might* be formed under certain conditions by volcanic activity—even though none has ever been found associated with volcanoes. The clay itself, in which the common clay mineral illite is replaced by smectite, *could* be altered volcanic ash. And rhenium, osmium, and iridium are known, though rarely, from volcanic eruptions.

There can be little doubt that vulcanism on the scale of the Deccan Traps might have brought about widespread destruction of living creatures on a local scale, but we have yet to see any mechanism by which vulcanism could have extinguished whole groups of animals on a worldwide scale.

The eruption of Mount Laki in Iceland in 1783 discharged about 12 cubic kilometers of basaltic lava (about equal to one decade's production of lava by the Deccan Traps); 75 percent of the country's livestock was killed and about 24 percent of the human population. Yet despite subsequent large eruptions in Iceland, the population, both human and wildlife, remains largely unaffected today; and there were no discernible effects beyond the shores of this small island. No single eruption within the Deccan Traps seems to have been much larger than the Laki eruption. Yes, such an eruption would cause massive deaths—locally—but not mass extinctions around the globe. All authorities are agreed (and I concur) that whatever the Cretaceous-Tertiary Boundary event was it killed 50 percent of families, 75 percent of species, and 99.99 percent of individuals—over the whole world.

Vulcanism Versus Impact

As I have stated above, the proponents of the volcanic theory of the Cretaceous-Tertiary event have pinpointed the Deccan Traps of India as the site of the major volcanic action they believe led to the great extinctions at this time, 65 million years ago. The Deccan Traps were certainly deposited by volcanic action at the right epoch and consisted of about a million cubic kilometers of volcanic scoria and lava, apparently deposited over a period of about a million years. Could this volcanic activity be considered sufficient to cause a catastrophic extinction of most of the species on earth?

Let's compare it with present-day volcanic activity. If the Deccan vulcanism deposited a million cubic kilometers in a million years, this amounts to an average of a hundred cubic kilometers per century. What is the sum total of volcanic discharge in this century? To take a few examples of recent activity, we have:

Pinatubo	2 cubic kilometers
El Chichón	2 cubic kilometers
Mount Saint Helens	1 cubic kilometer
Alaska in 1980s	2 cubic kilometers
Hekla (Iceland) 1947	1 cubic kilometer
Hawaiian volcanoes since 1950	17 cubic kilometers

Altogether, in the period 1900 to 1992—not even a full century—the total amount of volcanic lava, scoria, and ash from major eruptions alone amounts to 117 cubic kilometers. The total volume of lava and scoria produced by volcanic action this century, if we include minor eruptions and eruptions under the sea, is about 350 cubic kilometers, considerably more than the estimates for production by the Deccan Traps volcanoes.

In other words, the Deccan volcanism of 65 million years ago is no greater, and indeed perhaps less, than current volcanic activity—which is hardly catastrophic. The total release of energy is of the same order of magnitude as that released at the Yucatecan (yes, that's the correct adjectival form of Yucatán, in the Mayan tongue) impact, but instead of being released in a few seconds it was released over a period of perhaps a million years, at the same rate as energy is released by vulcanism this century.

The vulcanism of the Siberian Traps, which formed at the time of the terminal Permian event—the event that wiped out the Paleozoic fauna and flora, killing off about 95 percent of species—has been taken by some scientists to be the cause of that event. During this event the volcanoes produced even more lava (about 3.6 cubic kilometers) over a period of time that has been variously estimated as 600,000 to five million years. These figures lead to estimates varying from 70 to 600 cubic kilometers per century as the rate of production of the Siberian Traps volcanoes, figures that span the level of current volcanic activity.

Indeed, if we consider the Hawaiian archipelago alone, which has formed over a period of five million years, we note that its total mass of lava and scoria is about five million cubic kilometers—more than the Deccan Traps, more even than the Siberian Traps. Its mean rate of deposition over that five-million-year period, which continues today, has been 100 cubic kilometers a century. Does that mean that we are living through a mass extinction today? Of course we are. But it is a result of our own human activity—our profligate use of natural resources and our pollution—not of the formation of Hawaii by vulcanism.

Could the Deccan Traps have resulted from an impact? They could well have. Perhaps the comet that struck the earth 65 million years ago in Yucatán cracked the mantle of the earth and created a hot spot. The compression wave from such an impact can travel through the center of the Earth, while the shear wave is confined to the surface, because the liquid part of the core cannot propagate a shear wave at all, any more than any other liquid can. The shear wave, which was certainly powerful enough to be felt around the world, was confined to the mantle. Waves traveling in opposite directions would meet at the antipodean point and interfere constructively at that point with each other and with the compression wave. That is to say, at a point diametrically opposite to the point of impact, enormously destructive local forces would have been felt that could indeed have fractured the crust and created a major volcanic plume, perpetuated as a hot spot. The crucial spot for this to happen is the antipodean point from the impact site.

The site in Yucatán, which is the best candidate for a major impact at the Cretaceous-Tertiary Boundary, is located today at about 20°N 87°W, while the Deccan Traps occupy an area roughly 13°–20°N 75°–80°E. These two locations today are hardly antipodean. Let us not, however, neglect continental drift. During the Tertiary era, which continues to the present, India has been moving northwestward, pushing up the Himalayas in the process, while the Americas have been moving westward more slowly. At the time of the Cretaceous-Tertiary event the Yucatecan crater would have been approximately at 20°N 84°W and the center of the Deccan Traps at 21°S 95°E, almost exactly antipodean.

As I write this there is no proof, but I suggest that perhaps the Deccan vulcanism (and the Siberian vulcanism too at an earlier date) could be a result of a comet's impact. The earth's mantle cracked, perhaps at a preexisting weak point or hot spot, when the shear waves, coming in opposite directions from the Yucatecan impact, met on the other side of the world, and lava poured out—like what happens when you bite into a ripe tomato, spattering the juice all over your companion as the far side splits open. But this, of course, is pure speculation.

The Size of the Bolide

Everybody—or at least everybody who regards the impact theory as correct—says that the bolide was about 10 kilometers across. How did this figure come about? And is it correct?

The best estimates we have of the size of the bolide come from a consideration of the composition of the rocks of the Cretaceous-Tertiary Boundary, especially the nonterrestrial components. The first attempt at estimating the size of the bolide was made by the Alvarez group at Berkeley, who used the early measurements of iridium concentration in the Cretaceous-Tertiary Boundary rocks. Now, iridium is not strictly an extra-terrestrial component. All the evidence suggests that it is equally distributed throughout all parts of the solar system, planets and asteroids, gas giants and comets. Its ultimate origin was in supernovae, including the one that set off the shock wave that initiated the condensation of the cloud of gas to form the entire solar system, the sun and planets, asteroids and comets.

Iridium, a platinum-group element, is one of the heavier elements, being both large in atomic mass and also high in density as a metal, much heavier than iron. In fact, it is the densest element. Like platinum it is also much rarer than iron. And like platinum most of it has followed iron to the center of the earth, into the iron core, by gravitational differentiation while the earth was still molten. In a molten earth the heavier metallic elements sank down toward the center, leaving the lighter slag of the siliceous rocks floating on the surface to form the mantle. This differentiation has resulted in a crust that is seriously depleted in platinum-group elements as well as in such more common metals as iron and nickel. Indeed, the most heavily mined nickel deposits appear to be extraterrestrial, the result of asteroid impacts.

Any object in the solar system that has passed through a molten phase—other planets, asteroids, but not the moon—has undergone similar differentiation, with iridium, platinum, iron, and nickel sinking, under the influence of gravity, to form the core, leaving a metal-depleted surface layer. Nickel, platinum, iron, and iridium are all depleted in the surface layers of planets and asteroids. Correspondingly, their cores are enriched in these metals. Only those solar system objects that have never melted—in other words, comets—and so could not undergo gravitational differentiation into core and outer layers have primordial values of iridium, platinum, and other heavy metals. The outer layers of asteroids, those portions that form the stony meteorites, contain no more iridium and other platinum-group elements than does the earth's crust. Comets have the primordial concentration of iridium, while asteroid cores hold about twice the concentration found in comets.

The Alvarez group calculated the total amount of iridium that would be present in the Cretaceous-Tertiary Boundary rocks, on the assumption that this amount was the same at sites around the world. The concentration, expressed in mass per square centimeter, was certainly the same at the sites available to them, so this was a reasonable assumption. They then showed that this total amount of iridium could have been supplied by an asteroid, presumably a metallic asteroid fragment, of 10^{18} grams. This gave an excellent first approximation of the mass, which has not been altered significantly by the discovery of some twenty or more sites where the Cretaceous-Tertiary Boundary rocks have been exposed and analyzed.

But then the calculations begin to go a little awry. The Alvarez group assumed that the bolide was an asteroid, a most reasonable assumption given the state of knowledge at that time (1980). They therefore calculated that if the asteroid had the density of a stony meteorite, roughly 2.5 grams per milliliter, then the diameter of a spherical object of this mass would be a little over 9 kilometers (actually 9.14 kilometers). But, in fact, it is metallic meteorites that contain an excess of iridium. As I have shown above, stony meteorites have been depleted in this element by the same processes of heat and gravitational differentiation that have driven iridium from the earth's crust into the core. Nevertheless, this figure, rounded off to a diameter of 10 kilometers, has become enshrined in the literature as the diameter of the Cretaceous-Tertiary Boundary bolide.

A more correct calculation would be based on the density of a metallic meteorite, about 6.5 grams per milliliter. A spherical object of this density, of a mass of 10^{18} grams, would have a diameter of 6.65 kilometers. If the bolide really were an asteroid fragment, this would be a better approximation of its diameter, since it could only be a core fragment, not a fragment of the mantle of an asteroid.

A third calculation is possible, of course, one based on a mass of 10^{18} grams, a calculation relying on the rather poorly known density of comets. The best figure we have for this is 0.1 gram per milliliter. A comet with the same concentration of iridium as a metallic asteroid would then have a diameter of about 26.7 kilometers. But comets and metallic meteorites do not have the same concentration of iridium: comets have primordial values, while the metallic cores of asteroids are enriched about twofold over this figure. Accordingly, a comet would have to have twice the mass of a metallic meteorite to deliver the same load of iridium. Assuming a density of 0.1 gram per milliliter, the calculation for a comet is thus as follows:

Mass of comet	2×10^{18} grams
Diameter of comet	33.6 kilometers

This is not far from the upper dynamic limit of about 40 kilometers for comets, calculated both by Öpik and by myself by different methods. If, however, the density of cometary material is taken to be 1.0 gram per milliliter, a figure suggested more or less by default by at least one authority, then the diameter would be 15.63 kilometers.

To summarize this section, then, let me make the following points. First, the mass of a metallic asteroid required to deliver the amount of iridium observed at the Cretaceous-Tertiary Boundary would be about

10^{18} grams, and its diameter a little over 6.5 kilometers (not 10 kilometers). Second, a stony meteorite could not deliver the iridium load, since such objects have been depleted in this element. Finally, a comet would have to have about twice the mass of an asteroid core fragment to deliver the same load of iridium (about 2×10^{18} grams) and a diameter a little over 33 kilometers if the comet's density is 0.1 gram per milliliter, or 16 kilometers if its density is as high as 1.0 gram per milliliter.

Calculations I have made on the basis of amino acids and of diamonds found in Cretaceous-Tertiary Boundary rocks and in carbonaceous meteorites have led to a mass of 2×10^{18} grams for the mass, just as in the above calculation for a comet. Finding these materials in the rocks, however, is not compatible with any bolide that is either a stony meteorite or an iron meteorite; a carbonaceous meteorite is the only form of meteorite that could have delivered these materials. In other words, a comet (or a comet remnant or fragment) is the only possible delivery vehicle for amino acids and diamonds (which we will explore later), and even for silicon carbide, spinel, and chromite grains.

Even the simple mass and size calculations lead to the conclusion that the impactor at the Cretaceous-Tertiary Boundary was a comet, not an asteroid. It certainly cannot have been a fragment of the mantle of an asteroid—in other words a stony meteorite—for that contains no excess of iridium or of other platinum-group elements; neither can it have been a fragment of an asteroidal core, for this would not have released the amount of energy needed to account for the effects noted. Thus, only a cometary impact can account for all the features of the Cretaceous-Tertiary Boundary event.

7

Theories of the Periodicity
of Extinctions

Will the unicorn be willing to serve thee, or abide by thy crib?
Canst thou bind the unicorn with his band in the furrow? Or
will he harrow the valleys after thee? Wilt thou trust him, because his strength is great? —Job 39: 9–11

Periodicities in animals have gener-
ally been studied more in terms of diurnal or annual rhythms rather than
in periods of millions of years between extinctions. But the statistical
methods are the same—and far too subject to error and artifact.

Some forty years ago there was a great flurry of research into what were
called circadian rhythms in animals, after it was realized that, once separated from the rhythm of daylight and the diurnal changes in temperature,
certain physiological rhythms continued with a period of not exactly a day
but about a day (*circa diem* in Latin). Usually this circadian period was
more like 25 hours than 24, generally longer than a day. Unfortunately
many of the biologists investigating circadian rhythms were rather naive
in their use of statistics.

In the 1950s a group of American statisticians pointed out that in order
to study circadian rhythms properly we must ensure that the experimental
animals were totally isolated from any possibility of exposure to the natural environment. They therefore suggested that the only possible animal
for this purpose was the unicorn, a well-known but totally mythical animal that could never be exposed to diurnal rhythms since it had no real
existence. The unicorn's natural body functions were then represented by
a table of random numbers. Using the normal methods of statistical analysis used by students of circadian rhythm, they then showed that this purely

artificial construct would yield apparent periodicities, which were obviously pure artifacts. This became known as the "unicorn fallacy." The unicorn fallacy has raised its ugly head again in studies that have suggested that mass extinctions are periodic—the same old statistical methods, the same old artifacts, the same old unicorn fallacy.

To illustrate the point I have constructed the graph in Figure 7.1 entirely from random numbers generated by computer by what is known as a Monte Carlo simulation (named for the great casino at Monte Carlo and the simulation's dependence on chance). Graph A represents the actual fossil record of mass extinctions. The question mark indicates the Cambrian explosion, showing that we have no real measure of the size of this extinction, since animals at this period had no hard parts. Graph B is the Monte Carlo simulation, which has been constructed entirely using random numbers and hence shows no true periodicity, only stochastic, or chance, effects. A comparison of graphs A and B, however, shows that B seems to be at least as "periodic" as the actual record in A. Neither of them in fact is periodic; any apparent periodicity is simply an artifact of the bad use of statistics.

Four out of five Monte Carlo simulations suggest no evidence of periodicity, but one in five (like the one illustrated here) have a purely spurious periodicity when analyzed by current statistical methods, a periodicity at least as good as that ascribed to the real fossil record. Wilt thou trust the unicorn?

Since at least the Devonian period mass extinctions have occurred on average every 25 to 30 million years. There are two contending views about this: either the extinctions are periodic or they are not. My own view is that the interval between successive extinctions is purely stochastic—that is, the length of the interval is a matter of mere chance—and that there is no periodicity at all. I believe that the apparent periodicity is an artifact of statistical analysis—in other words, the unicorn fallacy.

Others have advanced several different arguments against any true periodicity, and it is my belief that not only are the mass extinctions of the Phanerozoic not periodic at all, but that they may well involve different mechanisms. It is likely enough that they all result from a bolide from outer space hitting the earth, but the nature of the bolide may differ from one to another event—a comet one time, an asteroid fragment the next—and the causation of the impact may differ too.

Despite all this evidence that mass extinctions occur at stochastic intervals, the belief that they are periodic has led to the elaboration of several theories to "explain" this "periodicity." All of them involve perturbation of the Oort Cloud of comets by external events. The three major hypotheses are as follows:

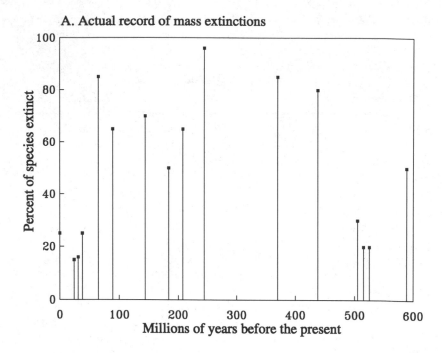

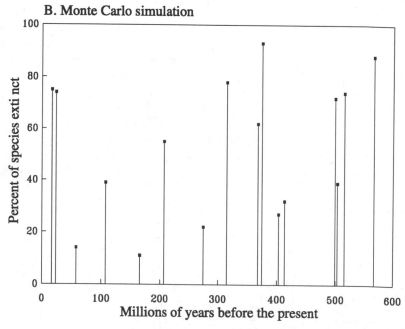

Fig. 7.1. A Monte Carlo simulation of the frequency of mass extinctions.

1. The sun has a companion star (forming a binary) in a highly eccentric orbit. This hypothetical star has been named Nemesis and is supposed to perturb the comets whenever it approaches the sun near enough, every 25 million years.

2. The sun, along with the whole solar system, is moving up and down like a carousel horse as it orbits around the galaxy. The central plane of the galaxy is known to harbor many giant molecular clouds. These are supposed to perturb the comets every time the solar system passes through the central plane, every 30 million years or so.

3. A hypothetical tenth planet (planet X) may perturb the Oort Cloud when it comes into conjunction with other planets, every 25 to 30 million years.

Nemesis

Once the idea of periodic extinctions had taken hold amongst some scientists, attempts were made to suggest cosmological mechanisms for such supposed periodicity. One of the first hypotheses was that of a companion star to the sun. Many stars we can see with the naked eye or with binoculars are binaries—that is, they are not single stars at all but pairs of stars, each orbiting the other.

Could the sun be a member of such a binary pair? Could a companion star to the sun provide the necessary periodicity for the supposed periodic extinctions? Such a hypothetical companion star has been named Nemesis, after the Greek goddess of vengeance. Nemesis is supposed to approach the sun closely every 25 to 30 million years, and when it does it perturbs the Oort Cloud of comets so that some fall in toward the sun, and the earth is exposed to a shower of comets that causes the extinctions.

Many papers and even one or two books have been devoted to the Nemesis hypothesis, which makes a fine scenario as long as it is presented only in words. But the moment we write down the equations that govern the motions of Nemesis, enter the necessary numbers, and solve them, difficulties arise. I do not intend to write down those equations here; instead I will content myself with a brief verbal account of the implications of this mathematical analysis. Other authorities have presented different criticisms of the Nemesis hypothesis, but my own are presented here.

What properties must Nemesis possess to cause repeated extinction events? First, it must pass near the sun once every 25 to 30 million years (depending on whose statistics you accept). Second, its mass must be much less than that of the sun and it must be fainter. The theoretical astronomy of Isaac Newton, and even earlier of Johannes Kepler, is enough to calculate the orbital characteristics of such a star as well as its mass, without having to bother with relativity; relativity calculations give much the same result. Such a star has an extremely elongated elliptical orbit, which mathematicians call highly eccentric. At its nearest approach to the sun it must pass within about 100 milliparsecs (about three light-months), within the Oort Cloud of comets but outside the outermost planets. At that point in its orbit Nemesis is supposed to perturb the comets in the Oort Cloud so that some of them start falling inward toward the sun, with one or more hitting the earth. At its most distant Nemesis is about 4 parsecs (about 13 light-years) away, three times the distance to the nearest star. At this distance it must be extraordinarily faint not to have been discovered by telescopic observation. Proponents of the Nemesis hypothesis seem to feel that it must be at the far end of its orbit just now, and that another extinction event is not due for another twelve million years.

It certainly seems possible that Nemesis exists, though it is perhaps strange that it has never been observed. It does not appear, however, to be a cause of major catastrophic events on earth, for several reasons. First, its orbit is unstable, since the distant parts of the orbit, beyond 1 parsec, are threaded between nearby stars. The perturbations created by these stars will cause a variation in the orbit so that the star will not return at the same closeness to the sun with each apparition; the length in time of different orbits may also vary by as much as four million years. But beyond this variability the orbit itself will decay after about 250 million years, and Nemesis will wander off on its own to other parts of the galaxy. There is no possibility that it could remain in such an unstable orbit around the sun for as long as even 300 million years.

Most binary pairs of stars are formed simultaneously by condensation from the same gas cloud. If the sun and Nemesis were formed in this way, they would have separated within 250 million years, before life even existed on earth. So Nemesis cannot have been part of the primordial solar system. That then leaves only the possibility that it was a passing star that was recently (within the last 250 million years) captured by the sun. But there is no known mechanism whereby this could occur. Gravity may drag a planet away from a star that passes too close—and indeed this may be the origin of Pluto—but it cannot account for the capture of a passing star. The earliest possible period during which Nemesis could have been captured for it to be still around today would be the Permian. What then could have been the cause of the major extinction events that occurred earlier, toward the end of the Devonian, the Ordovician, the Cambrian, and the Precambrian, the last leading to the Cambrian "explosion"? Logic seems to suggest that Nemesis simply does not exist.

But even if it does exist there is another powerful argument why it could not cause the extinctions: there simply have not been enough catastrophic events if Nemesis is the chief cause. We have abundant statistics on the mean frequency of closest approaches of passing stars, and it is quite clear that on average another star—a full-sized star, not just a dwarf like Nemesis—passes at least as close to the sun as Nemesis is supposed to do, but even more frequently. If the approach of Nemesis to within 100 milliparsecs were enough to cause a catastrophe like the Cretaceous-Tertiary Boundary event, then such a catastrophe should occur every time another star approaches this close (see Figure 11.1).

Finally, detailed computer modeling of the effects of the close approach of a passing star on the Oort Cloud of comets shows that though they would be perturbed, the perturbation would be outward, away from the sun and the passing star, not inward toward the sun.

An interesting corollary to this is the possibility that every normal star—every star in the same class as the sun, that is—possesses its own Oort Cloud. One of the problems in considering the evolution of the Oort Cloud is that in four and half billion years, so few comets have been lost to theft by passing stars. If each star has its own cloud, then at each close passage of such a star a few comets would be lost to interstellar space, but the passing stars would exchange a certain proportion of their comets. The present-day Oort Cloud thus may well include a mix of comets that were formed around many different stars, all at roughly the same time in a single star-forming event in a single star cluster. However that may be, it seems certain that Nemesis, if it even exists, which seems unlikely, cannot be the cause of repeated catastrophes like the Cretaceous-Tertiary Boundary event.

Giant Molecular Clouds

The solar system was formed initially from a giant molecular cloud by condensation of the gas. Such clouds are scattered in abundance throughout the galaxy, and many of them will sooner or later give birth not to single stars but to a star cluster as several centers of condensation form. One such cloud where active star formation is going on today is the nebula in Orion.

These clouds receive their name because the hydrogen, the chief component, is not ionized but in the form of molecular hydrogen, H_2. The total mass of gas in each one is enough to form a whole host of stars; indeed, when the sun was formed from one of these clouds it was only one of a whole cluster. If the solar system today were to pass through one of the clouds, the gravitational effects on the system of the proximity to such a mass of matter would be considerable.

Although giant molecular clouds occur throughout the galaxy they are most abundant in the central plane. Just as the moon orbits around the earth and the earth around the sun, so the sun (and with it the whole solar system) orbits around the galaxy. But unlike the earth, which keeps its orbit in a single flat plane, the plane of the ecliptic, the sun does not have a flat orbit around the galaxy but goes slowly up and down like a carousel horse. *Slowly* is the key word: the sun takes 200 million years, or thereabouts, to make one orbit of the galaxy, and the carousel horse takes 60 million years to rise and fall once.

Of course, any giant molecular cloud that is at the same distance as the sun from the center of the galaxy is also orbiting around the center at the same speed. Thus it is possible that the sun could be passing through the same giant molecular cloud every 30 million years or so, once on the way down and once on the way up, as the cloud (which is not bobbing up and down) keeps pace with it. At the moment ("moments" are rather long in this kind of context), the carousel horse is at the top of its ride. Any giant molecular cloud is most likely to be spread along the central plane of the galaxy, so it might be fifteen million years before the sun enters it again (if it ever has before).

The first problem with this hypothesis, then, is whether indeed there is a giant molecular cloud in the right position for the sun to enter it at regular intervals. We have lots of evidence for giant molecular clouds in general, since molecular hydrogen has a distinctive signal for the radiotelescope. But a cloud near enough for the sun to enter it every 30 million years should give a strong signal; yet no such nearby cloud has been found in the required position. We can plot the trajectory of the sun through the galaxy for the next billion years, but we have no evidence that a giant molecular cloud lies on this path.

But what would happen if the solar system did in fact enter a giant molecular cloud? The mass of the cloud would be perhaps a thousand times the mass of the sun, thus exerting enormous gravitational effect on the system. The relative speeds involved are quite small, not relativistic, so that Newtonian mathematics are good enough, at least for a first approximation, to calculate what will happen; there is no need to approach the calculations through relativity. In Newton's theory of gravity any mass can be treated as acting at a point, the center of gravity, thus simplifying the calculations. The calculations are simplified even more by treating the system in the framework of vortex theory, in 2.58 dimensions.

Entry of the solar system into such a cloud would exert a drag on all the planets, asteroids, and comets, perturbing their trajectories. But this perturbation is not such as to cause comets (or asteroids for that matter) to fall inward toward the sun and the earth. All the circulating bodies in the solar system would be slowed down in their orbits. The result of slowing down an orbiting body is to move it into a more distant orbit; in terms of artificial satellites orbiting the earth this would be equivalent to moving to a higher orbit. A shuttle 160 kilometers up takes some 90 minutes to orbit the earth, while a television satellite in geosynchronous orbit, near 50,000 kilometers up, makes one orbit in exactly 24 hours.

The drag of a giant molecular cloud would thus cause all orbiting bodies—planets, moons, asteroids, and comets—to move into "higher" orbits, farther from their primaries. The moon would move farther from the earth, the earth from the sun, and the Oort Cloud farther away. There is no indication that any orbiting objects would in fact start moving in toward the sun.

The orbit of the earth has been modeled by computer simulation for a billion years back, and there is no indication that this kind of perturbation has ever occurred. The hypothesis that a giant molecular cloud has perturbed the comets, has as a corollary the perturbation of the orbit of the earth, moving it every 30 million years progressively away from the sun. This has not happened. Nor has this kind of perturbation of the Oort Cloud of comets occurred at any time. A single perturbation of this kind, let alone one every 30 million years, would disperse the Oort Cloud into outer space and the solar system would by now be denuded of comets—and the earth would by now be orbiting out near Mars.

Planet X

Computations of the orbit of Neptune in the early years of this century indicated that another planet must be perturbing its trajectory. This theoretical finding led ultimately to the discovery of a ninth planet, Pluto. More refined calculations made after the discovery of Pluto suggested that it was not big enough to account for all the discrepancies in Neptune's orbit, leading some theoreticians to propose yet another planet, a tenth planet, or planet X (X is both the Latin numeral "ten" and also the symbol for the unknown). Though the existence of this planet was postulated only to account for the supposed erratic orbit of Neptune, it was soon seized on as a possible agent for causing perturbations in the Oort Cloud. The idea was that its conjunctions with gas giants could disturb the Oort Cloud in some way and cause a few comets to fall in toward the sun.

The planets orbit the sun at different rates according to their distance from the center of the solar system. Two planets are said to be in conjunction when their positions lie on a single straight line from the center of the sun or from the earth. In the latter case we see the two planets almost superimposed on one another in the sky. The outermost planets may take a hundred years or more to orbit the sun. A simple conjunction between any two of them occurs at an interval slightly longer than the shorter of the two orbital periods, just as there is "conjunction" between the two hands of a clock slightly less than once an hour—at noon, at about five minutes past one, ten minutes past two, and so on.

A conjunction between three planets occurs far less often, and a conjunction between four—say between Jupiter, Saturn, Neptune, and planet X—occurs only about once every 25 to 30 million years, depending on just where planet X is supposed to orbit. The coincidence with the period of this four-way conjunction and the supposed period of extinctions has led several authors to propose that this is what causes periodic perturbation of the Oort Cloud.

This hypothesis runs into several problems. First, there is no evidence that planet X even exists. More recent observations of the orbits of Neptune and Pluto suggest that any measured irregularities are within the limits of observational error and that there is no need to invoke a tenth planet to explain them away. Since this was the only reason to propose the existence of planet X, trouble arises when we try to invoke its baleful influence to wipe out the dinosaurs.

Second, even if planet X does exist, it must be much smaller than the gas giants. There is simply no room for another big planet outside Neptune, and any big planet would perturb Neptune far more than even the proponents of planet X would suggest. What then can a small planet (smaller even than Pluto; see Figure 5.3) do to the Oort Cloud that a three-way conjunction of the three gas giants cannot? These three-way conjunctions happen much more often than every 30 million years, so if such conjunctions really do affect the comets, then we should have far more mass extinctions than one every 30 million years.

Third, detailed modeling of the effects of planetary conjunctions on the Oort Cloud suggests that the one single result is a resonance in the middle of the Oort Cloud, like the Kirkwood gaps in the asteroid belt, dividing the cloud into an inner and outer Oort Cloud (see Figure 5.6). There is no mechanism whereby any planetary conjunction could cause comets to fall in toward the sun and earth. In other words, planet X could not have perturbed the Oort Cloud in such a way that it sends comets on a death mission toward the earth, and in any case planet X just does not even exist.

This seems to be a purely negative chapter, demolishing other peoples' theories. I have tried to show that mass extinctions are not at all periodic, but rather occur at stochastic or chance intervals. The average interval between extinctions is about 25 to 30 million years, but this is not real evidence of periodicity. Besides the evidence I have put forward, many other authors have advanced other reasons to think that there is no periodicity, but I have not summarized them here, preferring to put my trust in the unicorn.

As if this were not enough, I have gone on to offer my own particular criticisms of the three main theories that have been advanced to "explain" the supposed periodicity of mass extinctions. Once more, other authors have offered reasons why these three hypotheses won't work, but I have preferred to provide my own. It's not that I don't agree with the criticisms other authors have offered—I think they are well argued—but I might as well add my own pennyworth and throw in new arguments too.

The main point is that we need to search for another mechanism—one that happens on average every 30 million years or so—that might disrupt the even tenor of the outer solar system and throw some bolides off balance and into collision courses with the earth. That is really what ties this whole story together.

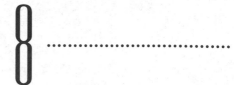

8

Meteors and Meteorites, Shooting Stars and Fireballs

Lucifer . . . shone like a meteor . . . in the streaming air.
—John Milton, *Paradise Lost*

Both meteors and meteorites are objects that enter earth's space from outside, and at first sight they seem to differ in little except size. This impression is quite wrong. Most meteors are cometary debris, whereas most meteorites, except carbonaceous meteorites, represent fragments of asteroids.

Meteors

A meteor is a falling star, or a shooting star, a small object that intercepted the orbit of the earth and was moving fast enough (in geocentric terms) to enter the atmosphere, where it burns up entirely, long before it could ever reach the surface. Some people use the word *meteor* to mean just the streak of light across the night sky, the shooting star as a visual event, just as John Milton did in the quotation that heads this chapter. Such people prefer the word *meteoroid* to denote the actual object that is burning up. I have never found this distinction useful; the context always determines the meaning. Besides, *meteoroid* is a wrong construction; the ending "-oid" in science always means "-like." A planetoid is something like a planet; ovoid means egglike or egg shaped. If a meteor is just a streak of light, a meteoroid can hardly be said to be like a streak of light, when it is being defined as an object that produces that streak of light. I refuse to use the term. A meteor remains a falling star and also the speck of matter whose passage causes the streak of light.

By definition a meteor never touches down. It generally burns up somewhere between 120 and 90 kilometers above the earth's surface. This means that we see the average meteor at a distance of up to 150 kilometers, since we will rarely see one directly overhead. I must admit that on the night when the Perseid meteor shower is expected to peak (supposing it is not raining) I usually go out onto my deck at home (I live in the country, far from city lights) and lie flat, wrapped in a sleeping bag, looking up at the stars. Even so I rarely see a meteor directly overhead, but generally somewhat to one side. (Yes, a warm sleeping bag at night in August—that's Canada for you.)

So a meteor is small enough to burn up completely in the upper atmosphere but releases enough energy to be visible at a range of 150 kilometers. The energy, of course, comes from the kinetic energy of the meteor's approach. Even so the amount released is astonishing. The typical meteor is between a microgram and a milligram—a mere speck of dust no larger than a salt grain. Yet the light trail is visible at 150 kilometers on a starlit night. Imagine the energy released by the arrival of something 10^{24} times bigger, which is the size of the object that caused the Cretaceous-Tertiary Boundary event.

What do we know about the composition of meteors? The major indications come from studies of the spectra of the shooting stars. The greater part of the light emitted is provided by components of the atmosphere, shocked by the passage of the meteor, but occasionally some spectral lines are distinct enough to give some idea of its composition. From these indications it seems that almost all meteors have more or less the same composition as carbonaceous meteorites or as comets, being rich in carbon and organic compounds. More recently this has been confirmed by satellite capture of extraterrestrial dust and direct analysis. Meteors in general, then, seem to be particles of carbon-rich dust.

Where do they come from? Most meteors come in regular annual showers, the Leonids for example. Such showers are rarely seen before midnight, usually peaking around two or three o'clock in the morning. This is because in the daily rotation of the earth, the afternoon and evening sky is that part of the globe that is rotating away from the direction from which the meteor swarm is approaching, while the late-night-to-early-morning side is moving toward the swarm. Few meteors are able to hit the earth on the afternoon-to-evening "retreating" side.

Radar observations show that meteor showers also occur during the daytime, being just as regular in their annual cycle as the familiar nighttime showers and always peaking before noon, for the same reason that nighttime meteor showers peak after midnight. Several such annual showers, both daytime and nighttime, have been identified with the trajectories of comets, and it is evident that the passage of a comet leaves a trail of dust along its orbit. A meteor shower occurs when the earth, in its annual passage around the sun, passes through the dust trail of a comet that may have passed this way centuries before and may even have disintegrated several thousand years ago. Indeed, the progenitor of the largest meteor stream, the Taurids of the southern hemisphere, appears to have been a comet that disintegrated at the beginning of the third millennium B.C. Thus, meteors, or almost all meteors, are cometary debris. In contrast, most meteorites, except carbonaceous meteorites, are asteroidal debris of varying chemical composition.

Fireballs

Some larger objects than shooting stars should also be considered as meteors, insofar as they never penetrate to the surface of the earth but blow apart in the atmosphere. These are generally known as fireballs. Anything, however fragile, larger than about a kilometer across will penetrate through the atmosphere to hit the surface of the earth, and many smaller objects too will do so, to become meteorites. But a fragile object less than a kilometer across will penetrate some distance into the atmosphere and then blow apart as a fireball. How far it will penetrate depends chiefly on its tensile strength, not on its size, if it is below a kilometer.

A fragment of the core of an asteroid, if it is big enough not to burn up completely in the atmosphere, will not end as a fireball; it has a high enough tensile strength to penetrate the full depth of the atmosphere to the surface as an iron meteorite.

A piece of the mantle of an asteroid may be strong enough to come down to earth as a stony meteorite, but if it is crumbly stone, fragile and friable, it may end as a fireball some 10 to 20 kilometers up in the air. The most spectacular such fireball of the twentieth century seems to have been the Tunguska event, in which a fireball exploding at about 10 or 11 kilometers up, devastated tens of thousands of square kilometers of the Siberian tundra in 1908. The incoming bolide seems to have been about 100 meters across.

Larger bolides than this have formed fireballs in this century without doing any significant damage, and these appear to have been cometary fragments that have burst into fireballs at a much greater height in the atmosphere, more than 50 kilometers up, an elevation so great that the tenuous atmosphere would not transmit a shock wave to the surface of the earth.

Of course some fireballs are not fully consumed; their fragments actually land as meteorite showers. A famous example is the huge fireball (apparently a stony object) that swept across New England in December 1807, crashing to earth as numerous fragments near the town of Weston, Connecticut. Benjamin Silliman (a Yale scholar who founded *Silliman's Journal*, a famous scientific periodical that continues publishing to the present day), together with a colleague, collected numerous meteorites from this fireball, including one weighing almost 100 kilograms. This event is said (perhaps apocryphally) to have prompted President Thomas Jefferson to remark, "It is easier to believe that two Yankee professors would lie than that stones would fall from heaven."

Meteorites

The word *meteorite* is somewhat ambiguous, since it is used in two different but related ways. It is more often taken to mean an object from outer space, an impactor or bolide, that is slowed enough by friction from the atmosphere so that although it hits the surface of the earth it does not make any kind of crater. When *meteorite* is used in this sense, it leaves us no really unambiguous term for a crater-forming bolide or impactor. The word is also used to mean any impacting object that actually touches down, whether or not it makes a crater. In this sense the impactors that created the craters on the moon are meteorites. In the former sense of the word they are not meteorites slowed by friction, for the moon possesses no atmosphere to slow down impactors. Take your choice. One of the problems in reading about science is often the frustrating problem of deciding the sense in which an author uses a word. Scientists seem to be worse than most other people in this regard.

The Composition of Meteorites

The vast majority of meteorites recovered on the earth's surface turn out to be stony objects formed of rocks whose chemical composition is like those on earth, chiefly silicates. Other meteorites are largely metallic, consisting chiefly of iron or nickel, together with other siderophilic elements. A few are carbonaceous in composition, containing up to 10 percent carbon, many organic compounds, and even diamonds. A very few (some so-called achondritic meteorites) are thought to be debris thrown up from the surface of Mars when a large bolide hit that planet. They share the composition of Martian rocks—not very different from terrestrial rocks but still quite distinctive.

The abundant stony meteorites differ from terrestrial rocks not so much in chemical composition but in the way in which the constituents are organized into what look like solidified droplets cemented or agglomerated together. These chondrules, as they are called, seem to represent primordial material condensed from the gaseous cloud at the time of formation of the solar system. The time of their formation has been dated by several techniques of radiometric dating, and uniformly their age is that of the solar system. The date these chondrules were cemented together is somewhat later, perhaps 4.52 billion years ago rather than the 4.54 billion years of the date of their formation.

The second most abundant meteorites are also chondritic (that is to say formed of chondrules), but in this case the chondrules are metallic, chiefly iron and nickel. The date at which they were annealed is somewhat later than the stony chondritic meteorites, probably when they were incorporated in the centers of asteroids.

There are a few achondritic meteorites (meteorites that lack chondrules) that during their history have been melted, thus completely destroying the chondrules. Most of these have a radiometric date of one to two billion years. Included among these are the meteorites of Martian origin. The date presumably reflects the time when they were remelted and thrown clear of the parent planet.

Finally, the carbonaceous chondritic meteorites are believed to be the oldest datable objects in the solar system (dated to 4.64 billion years old) and are the focus of some interesting studies and speculation. Since some of them are slightly older than the solar system, I believe they come from other older stars (in the same star cluster) captured during a close approach.

The stony and the iron chondritic meteorites (and a few intermediate types) are certainly derived from asteroids, broken up by collisions induced by resonance with the gravitational field of Jupiter. Achondritic meteorites appear chiefly to be ejecta from meteoritic impacts on the moon or on Mars. The origin of the carbonaceous chondritic meteorites is a subject of dispute; many people think that they too are of asteroidal origin. This I doubt. Let me explain why.

Carbonaceous Meteorites and Comets

Asteroids were formed in the inner zone of the solar system under conditions of excess oxygen. Accordingly, they consist primarily of siliceous rock and quartz, rocks in which silicon has combined with excess oxygen. Silicon in carbonaceous meteorites is usually in the form of silicon carbide, as it is in comets, formed in the outer oxygen-starved part of the solar system. Free carbon seems rare in asteroids and when it does occur the isotope ratio is about 90:1, the value characteristic of the stony planets. Up to 5 percent of a carbonaceous meteorite may be in the form of free carbon (with an isotope ratio of 42:1); 20 percent of that is in the form of diamonds. The carbonaceous meteorites were formed earlier than the stony ones; the outer zone of the solar system (including the gas giants and the comets) formed before the inner zone. The inner zone of the solar system, including the earth and the asteroids, is enriched in carbon 12 compared to the outer zone and to interstellar dust. The carbon of carbonaceous meteorites has the carbon 12 levels of interstellar dust, not that of the inner solar system. All this adds up to a cometary origin for the carbonaceous meteorites.

This is not to say that they have arrived on earth directly from the Oort Cloud of the outer solar system. Perturbation of cometary trajectories will often result in comets being parked in the asteroid belt; many small asteroids show the surface brightness not of light-colored rocks but of carbonaceous cometary material. A fragment of a comet parked in orbit in the asteroid belt would be baked by the sun, driving off all volatile materials, which are the principal components of comets, and leaving only the non-volatile components. Such material is said to be a "dead" comet. Dead comets in the asteroid belt are just as subject as primordial asteroids to perturbation and fragmentation by the gravitational field of Jupiter.

I conclude that carbonaceous chondritic meteorites represent fragments of dead comets, while the much more abundant noncarbonaceous chondritic meteorites are fragments of asteroids. By contrast, meteors represent cometary dust. Another way of putting it is that the death of a comet may give rise to the formation of a carbonaceous asteroid. Comets repeatedly passing the sun boil off volatile materials at each apparition and may ultimately be trapped by resonance with Jupiter in the asteroid belt. Here they would be subject to final depletion of their volatiles by the sun. Öpik has suggested that inactive comets might be masquerading as earth-crossing asteroids, while the outermost known asteroid, named 2060 Chiron, shows suspiciously cometlike behavior, including the formation of a coma (a sort of gas cloud) at its closest approach to the sun. Fred Whipple of Harvard University (who first suggested that comets were like dirty snowballs) has pointed out that the Geminid meteor stream and the asteroid 3200 Phaeton have identical orbits, suggesting that Phaeton is an extinct or dead comet that at some time in the past, before its death (that is to say before it lost its volatile materials), shed the stream of dust that now forms the Geminids.

Indeed, we have to consider what happens to comets as they near the end of their life in the inner solar system, baked by the sun, their volatile components evaporating. Paul Weissman has calculated that 65 percent of all comets entering the inner solar system disappear because they are ejected as a result of strong perturbation by the gravity of Jupiter, producing truly interstellar comets.

About 27 percent of inner solar system comets are disrupted by the tidal effects of close approaches to the sun or to one of the gas giants, forming rivers of dust. Disruption of comets in this way has actually been observed; comet Biela was seen to break up first into four pieces in 1852 and has not been seen since; comet Brorsen 1879, comet Westphal 1913, and comet Neujmin 1927 are other examples of comets actually seen breaking apart and then never seen again.

We have so far accounted for the disappearance of 92 percent of the comets that enter the inner solar system—the so-called short-period comets. A few of the remainder commit suicide by hitting a planet or a satellite, but Weissman has calculated that 7 percent are perturbed by resonance with the gravity of Jupiter to end their days in the asteroid belt. Whether we then regard them as asteroids or not depends on our definition. If we regard any small object orbiting the sun in the asteroid belt (other than an active comet) to be an asteroid, then they are dead comets that have evolved into carbonaceous asteroids. If, however, we define an asteroid as an interplanetary body that formed without appreciable ice content (the definition proposed in 1987 by William F. Hartmann of the University of Arizona), then a dead comet, even if it circulates in the asteroid belt, can never be called an asteroid. What then do we call carbonaceous objects that circulate there?

This discussion of meteorites includes both meteorites in the narrower sense of meaning—those objects small enough to be slowed down by the drag of the atmosphere to such an extent that they do not form craters—as well as larger impactors that form craters. Like meteorites in the narrow sense, these larger impactors were probably mainly asteroid fragments, mostly stony chondritic objects. But some of them could also have been entire comets not more than about 40 kilometers across. Certainly the impactor that formed the Barringer Crater (also known as the Meteor Crater) in northern Arizona some tens of thousands of years ago was a metallic object about 42 meters across, while the Sudbury impactor in northern Ontario, Canada, that hit the earth about 1.82 billion years ago was a metallic meteorite considerably larger than 40 kilometers. The Tunguska event in Siberia in 1908 was a small fragile stony meteorite fragment perhaps 50 or 100 meters across that blew apart to form a fireball at an elevation of about 10 or 11 kilometers above the surface of the earth (not a "dead" cometary fragment), and I hope in the next chapter to show that the Cretaceous-Tertiary Boundary event involved a cometary impact.

Shooting Stars

Since we are observing it at a range of 100 to 150 kilometers, it can hardly be the object itself that is shining white hot. Certainly we can see a space shuttle at that sort of range when it is still illuminated by the sun just after sunset, but we should have trouble seeing an object as small as a pickup truck with the naked eye at that distance, even if it were white hot. A stony meteorite the size of a pickup truck would have a mass of about 20 tons, so that gives some idea of the smallest meteor we might observe with the naked eye—if we were actually observing it by its own incandescence. But something the size of a grain of salt? What we are actually seeing is not the dust particle itself (what some people call the "meteoroid," though I won't) but the trail of ionized air molecules created by the shock wave—the sonic boom—of its passage.

The average meteor enters the atmosphere at a speed of about 18 kilometers per second. When put like this, such a speed is difficult to comprehend. A car traveling along the interstate at a little over 55 miles per hour is moving at 100 kilometers per hour, which works out to about 28 meters per second. A Boeing 747 cruising at its normal 550 knots is doing about 300 meters per second, while the Concorde, at Mach 2, is traveling at about 700 meters per second. The speediest aircraft in the U.S. Air Force (at least in the public domain) can reach Mach 3 at 80,000 feet (25,000 meters). It is traveling at a little over 1 kilometer per second, at which speed the titanium skin becomes an incandescent red hot. The average meteor is burning up and creating a compression wave at eighteen times that speed, or 650 times the speed of a car on the highway.

I have had personal experience of the energy of a compression wave when I was a little too near an underwater mine that exploded, killing my diving buddy. We were about 25 to 28 meters down when he made a mistake and triggered the mine. They never did recover any piece of him. I was more fortunate; I was about 15 meters away. The compression wave from the blast tore my face mask and my mouthpiece off, taking a few teeth with it. My eardrums burst, the lenses of my eyes developed cavitation bubbles, and I was twisted into a pretzel. I lost consciousness, and what saved my life was a piece of shrapnel that cut my weight belt, releasing about 10 kilograms of lead. I bobbed up to the surface and was rescued by the diving boat, bleeding like a stuck pig. (What a cliché! But that's what happens when you're gashed in the belly by shrapnel.)

It is simple to say that the injuries were the result of the blast, but how did the blast cause the injuries? The mechanism of transfer of energy from the high explosive to my body was a compression wave. Just the same kind of compression wave is what raises the atmospheric gases to incandescence and creates the visible streak of light. This is the mechanism by which the kinetic energy of the incoming dust particle is converted into visible light; a compression wave is the mediating mechanism for the conversion of kinetic energy, the energy of motion, into light energy and heat.

We can readily see the meteoric trail even if we cannot hear the sonic boom of its supersonic passage.

The sonic boom from an object, even a tiny object, moving at the sort of speed achieved by an incoming meteor, is enough to compress and hence heat the gases in the air along the trail sufficiently for oxygen and nitrogen to separate into their constituent atoms, or rather ions, forming a hot plasma. This is what we see as a shooting star—not the dust particle itself, but the very atmosphere "burning" in its trail.

When the ionized atoms of oxygen and nitrogen recombine again, most of them pair up with their like, but a small percentage actually pair a nitrogen atom with an oxygen atom, forming oxides of nitrogen. These ultimately react with water to form nitric acid. Thus an ultimate result of a meteor—or of a fireball or meteorite—is acid rain.

9

What Hit at the Cretaceous-Tertiary Boundary

And the fifth angel sounded, and I saw a star fall from heaven unto the earth: and to him was given the key of the bottomless pit. And he opened the bottomless pit; and there arose a smoke out of the pit, as the smoke of a great furnace; and the sun and the air were darkened. —Revelation 9: 1–2

I have argued in previous chapters that "a star fell from heaven unto the earth" (a bolide of some kind), that it created a "bottomless pit" (a crater), and that "there arose a smoke out of the pit . . . and the sun and the air were darkened" (a cloud of ejecta was formed from the crater that spread around the earth in the stratosphere and later precipitated to form the Cretaceous-Tertiary Boundary layer in the rocks). I can hardly think of a better poetic description of the Cretaceous-Tertiary Boundary event than that of the third-century mystic whose vision is quoted above.

The original hypothesis advanced by Luis and Walter Alvarez and their coworkers in 1980 was that an asteroid hit the earth at the time of the Cretaceous-Tertiary Boundary. More recently it has often been tacitly assumed that the impactor was a comet rather than an asteroid or that it was a veritable shower of comets, as proposed by the Dutch astronomer Piet Hut. Which was it? In order to answer this question we must first review what is known about the chemical composition of both these types of satellites of the sun.

Comets and Asteroids

Asteroids are components of the innermost region of the solar system, the zone of the stony planets, and share with those planets a number of features. First, this zone was formed under conditions of oxygen excess, or, more specifically, in a region where, atom for atom, oxygen greatly exceeded carbon. As a result (except where living organisms have wreaked their usual havoc), most carbon is oxidized (as carbon dioxide or carbonate); so too is all silicon oxidized to form the silicate rocks, the predominant rocks of earth, of the stony planets in general, and of the asteroids.

Second, the asteroids, like the stony planets, have been subjected to enough heating to melt them at some time so that the heavy siderophilic elements (iron, nickel, cobalt, and the platinum-group elements) have separated out as a distinct heavy core with the silicate rocks forming a crust on the surface. Like the planets, this stony crust (the source of stony meteorites) is depleted in the platinum-group elements, while the core (the source of iron meteorites) is enriched. This thermal differentiation has also resulted in the destruction of some components, such as diamonds, which we have reason to believe were a primordial constituent of the stuff from which the entire solar system was made.

Because of the disruption of orbits in the asteroid belt caused by the gravity of Jupiter and by resonance effects, no asteroid existing today can be a whole planetesimal. Repeated collisions have smashed them apart and recombined them. As a result, any asteroid that leaves the asteroid belt to impact with earth is likely to be predominantly either a fragment of crust or a fragment of core.

Certainly such objects have hit the earth in the past, including most meteorites, which are fragments of asteroids. Indeed, the largest such object of which we have a geological record was the object, more than 100 kilometers across, that created Canada's Sudbury Basin. Could such an object, some 6 to 10 kilometers across, be the impactor at the Cretaceous-Tertiary Boundary?

At first glance this could certainly be so. We must, however, place certain constraints on its type. The major point made by the seminal 1980 paper by Alvarez and his group was the anomalously high levels of several platinum-group elements, especially iridium, at the Cretaceous-Tertiary Boundary. Indeed, the phrase "iridium anomaly" has entered the geological vocabulary to denote any extinction boundary. These elements— iridium and other platinum-group elements—are enriched in asteroid core material and depleted in asteroid crustal material, so the impactor, if indeed it was an asteroid fragment, must have been a core fragment—that is, it must have partaken of the nature of an iron meteorite. The problem

TABLE 9.1

Summary of Differences in the Composition of Comets and of Asteroids

Composition	Asteroids	Comets
Volatiles	Absent	Abundant
Thermally differentiated	Yes	No
Silicon	Silicates	Some silicates
Silicon carbide[a]	No	Yes
Free carbon	No	Graphite, diamonds[a]
Oxidized carbon	Abundant	Rare
Organic carbon	No	Abundant; amino acids[a]
Siderophilic elements	Crust depleted; core enriched twofold	Primordial abundance
Chromite, spinel[a]	No	Abundant
Carbon isotopes[a]	^{12}C enriched (like the earth)	Not enriched in ^{12}C

[a] Items on which I have concentrated my research.

is, we have no other evidence besides the iridium anomaly that it was in deed of this nature.

In contrast to the asteroids, comets were formed in the outer solar system, beyond the planets, a region where oxygen is (and was) relatively deficient. More precisely, atom for atom, there is less oxygen than carbon. All the available oxygen (or most of it) was used up in reacting with silicon to form silicates. And still there was silicon left over, while almost no oxygen was free to combine with carbon. Accordingly, comets contain much free carbon, almost unknown in the planets and asteroids of the inner solar system, except where living organisms have interfered with nature. Moreover, carbon and silicon have combined together to form silicon carbide grains, which are never found in stony or iron meteorites.

As a final distinction from asteroids, comets have never been subjected to excessive heating and so contain the volatile materials missing from asteroids; comets have never undergone differentiation into metallic core and rocky mantle. Instead, their constituent components are present in the primordial proportions in which they formed initially from the nebular cloud of the presolar system. The differences between asteroids and comets are summarized in Table 9.1. With this table we can begin to design a search for constituents of the Cretaceous-Tertiary Boundary rocks that might enable us to decide on the nature of the impactor.

Spinel and Chromite

The presence of spinel and chromite is by no means diagnostic of cometary material (or, more precisely, of carbonaceous meteoritic material) but is perhaps suggestive, especially since they have not been reported as constituents of stony or iron meteorites. Both have the same crystal form;

indeed, chromite can be regarded as a form of spinel in which aluminum has been replaced by chromium. Both are hard refractory oxide minerals that withstand dissolution in strong acids, even in hydrofluoric acid, which will dissolve glass and silicate rocks. Both of them are found on earth; some forms of spinel are semiprecious gemstones, notably the red spinel ruby or balas ruby, while the only commercial chromium ores take the form of chromite.

If we take a few grams of Cretaceous-Tertiary Boundary rock and dissolve the major components in acid, what results? Which acid we use will depend on the nature of the rock. The limestone of the Gubbio section in Italy dissolves readily in hydrochloric acid; the claystone of the Canadian Red Deer Valley will only dissolve in hydrofluoric acid. In either case the undissolved material can be centrifuged off. Any remaining silicates in the Gubbio rock can then be dissolved by washing with hydrofluoric acid, and coal particles can be removed by treating with nitric acid and perchloric acid. What remains after all these washings is a few milligrams of grains of material that will not dissolve in any of the acids we have used.

In all the Cretaceous-Tertiary Boundary rocks I have treated this way, the bulk of these grains are always spinel and chromite, in roughly equal proportions. The identification must be made under the microscope or by other instrumental means, but it is unequivocal. In addition, individual hand-picked chromite grains can be analyzed chemically and shown to consist largely of the element chromium; the grains contain almost 50 percent of this element.

But this is only true of the Cretaceous-Tertiary Boundary rocks themselves. Samples taken 2 centimeters above or below the boundary, or 5 meters above or below, contain no detectable spinel or chromite grains, and these grains have not been reported in most bedrock, only in some igneous mafic ores.

In the sedimentary rocks I have studied, spinel and chromite are characteristic of the Cretaceous-Tertiary Boundary layer and are not found in other rocks. They are characteristic of comets (and of carbonaceous meteorites, which ultimately derive from comets) and are not found in stony meteorites or metallic meteorites; hence they are not common asteroidal constituents.

Carbon Isotope Ratios

All isotopes of the same element have the same number of protons in the atomic nucleus; this proton number is what gives rise to the identity of the element and to its unique chemical properties. Different isotopes of the same element differ in the number of neutrons contained in the nucleus. Carbon has six protons in its nucleus but may have from six to eight neutrons. Protons and neutrons together are called nucleons. Carbon 12, the most common form of carbon, has twelve nucleons, six protons and six neutrons; the rarer carbon 13 has seven neutrons, while carbon 14, which is unstable and therefore radioactive, has eight neutrons, making fourteen nucleons in all. In considering the Cretaceous-Tertiary Boundary, we are concerned only with carbon 12 and carbon 13, which contain twelve and thirteen nucleons, respectively.

In the general neighborhood of the sun and the solar system, the proportion of carbon 12 to carbon 13 is about 42:1. This ratio is known from astronomical studies both of nearby stars and of interstellar dust, as well as from recent satellite flybys of comet P/Halley and other comets. The same ratio is found in the atmosphere of Jupiter and the other gas giants as well as in the sun itself.

The situation is far different for the inner stony planets and the asteroids. For some reason as yet unexplained in detail, though probably having to do with the initial formation of the planets, these objects are enriched in carbon 12. Terrestrial carbon, as well as carbon on the moon, Venus, and Mars, has about 89 to 92 times as much carbon 12 as carbon 13. This same ratio is found in asteroidal fragments, stony and metallic meteorites, and the spectra of normal asteroids. The sole exception is the carbonaceous meteorites, derived from cometary fragments, which have the interstellar ratio of the two carbon isotopes, roughly 42:1.

What then of the carbon in the Cretaceous-Tertiary Boundary rocks? This question is not quite as simple as it may appear, for the Cretaceous-Tertiary Boundary material is only about one part of impactor to twelve parts of material ejected from the crater. Thus, it becomes necessary to separate constituents that are of undeniably extraterrestrial origin. I will therefore return to this question later.

Diamonds

Free carbon can exist in three entirely different forms: graphite, diamond, and buckminsterfullerene (commonly known as bucky balls). These forms differ in their crystalline structure, and under certain conditions they can be converted one into the other. Under terrestrial conditions graphite is the most stable of the three forms, being black, soft, and almost soapy to the touch. Bucky balls are less common but are easily prepared in the laboratory; in an impure form they are common constituents of soot. Least common are diamonds, which are hard, white (at least in the purer forms), and stable except when heated to high temperatures, when they will burn just as readily as charcoal.

Terrestrial diamonds are the result of conversion of other forms of carbon (probably chiefly methane, not free carbon) into crystalline form under prolonged heat and pressure in the roots of volcanoes. Most diamonds of commerce were formed over a billion years ago, many more than two billion years ago, and have been brought to the surface by the gradual elevation of these volcanic roots and the erosion of the rocks overlying them. Roughly put, the formerly molten lava pipes, the bases of the conduits by which magma was once brought to the surface a couple of billion years in the past, are now the kimberlite pipes, the mother lode of diamonds. Needless to say, these diamonds share the carbon-12 enrichment of mother Earth. In the way that the carbon isotope ratio is usually measured, we say that the $\delta^{13}C$ (carbon-13 deficit) has a mean of about -7 ‰ (parts per thousand), the same as the carbonate of terrestrial limestone. Figure 9.1 shows the carbon-13 deficit for meteoritic, terrestrial, and Cretaceous-Tertiary Boundary diamonds, and it is obvious that the first and last of these three agree, being quite different from terrestrial diamonds.

But there is another way in which diamonds can be formed, as discovered in the attempts to make diamond films for semiconductor chips—namely, under low pressure and temperature. Organic compounds in the presence of a large excess of hydrogen, say one part of methane to fourteen parts of hydrogen, when subjected to a suitable energy flux, can form diamond instead of the expected graphite. In the laboratory, such diamond is deposited on a substrate to form a film. Thermodynamic calculations suggest that under some conditions this diamond material may be more stable than graphite, while "normal" diamonds are not actually truly stable, but rather metastable.

Conditions around a supernova, soon after it has blown its top, are ideal for the formation of diamond by just this mechanism, and some calculations suggest that as much as 50 percent of the carbon blown off by a dying star when it turns supernova may end up as diamond rather than as graphite. Indeed, it seems that any dying star, once it has begun to outgas in its death throes, may develop a halo of diamonds, formed in the stellar wind or cloud of hydrogen tinged with carbon. This is true whether we are dealing with a red giant, a nova, or a supernova; the process does not require the high energy of the shell of debris that surrounds a supernova. Once formed these diamonds are stable. Around a supernova or any other dying star, of course, there are no surfaces on which the diamond may form a film, and the stable form of diamond under these circumstances is

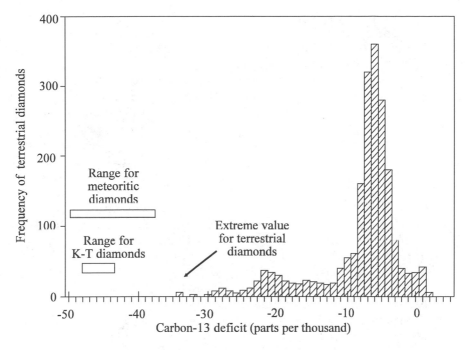

Fig. 9.1. The carbon isotope ratios in diamonds. An internationally defined sample of limestone is the arbitrary standard against which all samples are compared, and by definition this has a carbon-13 deficit of zero. The mean value for sedimentary rocks is a deficit of seven parts per thousand, written as − 7‰. Terrestrial diamonds too mostly have a deficit of − 7‰, which corresponds to a carbon-12-to-carbon-13 ratio of about 90∶1. Spectroscopic studies of the tails of comets and of interstellar dust have shown that the more usual ratio for these two isotopes of carbon in the nearer parts of interstellar space and in comets is about 42∶1, so that the earth and the inner solar system generally are somehow enriched in carbon 12. Diamonds from meteorites and those from the Cretaceous-Tertiary Boundary claystone show the carbon isotope ratio characteristic of interstellar dust, with a carbon-13 deficit of about − 45 to − 50‰. This corresponds to a carbon-12-to-carbon-13 ratio of about 42∶1. (The same is true of the amino acids found in these two sites.) Updated from a diagram published by Carlisle (1992), incorporating data (on the right side of the figure) provided by Galimov (1991).

as submicroscopic dust particles about 3 to 5 nanometers across, containing no more than 20,000 to 50,000 atoms each.

How tiny is that? If we were to string a thousand of these diamonds together to make a necklace, it would just about fit a bacterium if a bacterium had a neck. Put another way, it takes many billion billions of these diamonds to make one carat; they are too small to see under the light microscope and are even difficult to image with the scanning electron microscope. Recent studies of interstellar dust in the vicinity of the solar system suggest that half the dust particles may consist of these minidiamonds or of clumps of them stuck together.

What has this got to do with the Cretaceous-Tertiary Boundary? The solar system was formed from a second- or third-generation nebular cloud containing all kinds of materials from earlier stars (including novae and supernovae). Condensation of this cloud began when a shock wave from a nearby supernova started a collapse to form a star cluster, including the sun, planets, and comets, as discussed in Chapter 5. Accordingly, the primordial material of the solar system included large amounts of this diamond dust.

In the inner solar system, the region of the stony planets, these diamonds were early destroyed by the gravitational heating that resulted in thermal differentiation into core and crust. No such diamonds are known in terrestrial rocks or in asteroidal fragments, stony meteorites, or iron meteorites. Comets, however, have never undergone this degree of heating. Even when their volatile fractions are driven off and they have been baked to a crisp by proximity to the sun, they have never been heated enough to destroy the primordial supernova diamonds. Accordingly, comets, and their derivatives, the carbonaceous meteorites, contain abundant nanometer-sized diamonds, formed around supernovae or other dying stars, primordial diamonds as old as or older than the solar system, at least 4.56 billion years old.

Needless to say, these diamonds, which we can extract from carbonaceous meteorites (cometary fragments), have a carbon isotope ratio characteristic of interstellar dust—namely, about 42 times as much carbon 12 as carbon 13. In some carbonaceous meteorites they make up as much as 5 percent of the total mass of material; in others, about 5 to 50 percent of the total carbon. They are unique to the carbonaceous chondritic meteorites and have never been found in stony or metallic meteorites. In other words, they are diagnostic of cometary material and absent from asteroidal material.

My colleague Dennis Braman and I have conducted a search of Cretaceous-Tertiary Boundary rocks for these diamonds. We have found them in abundance in the Cretaceous-Tertiary Boundary rocks of Alberta. "Abundance" is a relative description, of course; the cost per carat of extracting them far exceeds the cost of commercial diamond mining. What is in fact important in this regard is that the ratio of the mass of diamonds recovered to iridium is almost identical in the rocks as it is in the best-known carbonaceous meteorite. In other words, if a bolide of roughly this composition delivered the iridium that forms the iridium anomaly at the

Cretaceous-Tertiary Boundary, then it would also have delivered the amount of diamonds we find in the rocks.

We found the diamonds by dissolving everything else with acids, except the diamonds, spinel, and chromite. Under the acid conditions we used, starting with a mere 20 grams of rock (less than an ounce), we were able to extract the spinel and the chromite, with the diamonds sticking to the surface of these two mineral grains, a process known as adsorption. The diamonds themselves were too tiny to handle on their own in impure form, but the larger (though still microscopic) grains of these other two minerals could be centrifuged to separate them from impurities. So long as the washing fluid remained acid, the diamonds were adsorbed onto the surface of these larger grains and did not wash off. Once we had got rid of everything else it was possible to wash the diamonds off the other mineral grains with slightly alkaline water and then evaporate off the water to leave a film of diamond crystals for analysis. They were then easily identified by electron diffraction spectroscopy and their carbon isotope ratios measured by mass spectrometry. As usual in this work, "easily" is also a relative term; the measurements were pushing the limits of sensitivity of available techniques, but the results were quite unequivocal. The tiny crystals proved to be identical with diamonds extracted from carbonaceous meteorites in size, crystal form, and carbon isotope ratio, which was 42: 1 (see Figure 9.1).

As with spinel and chromite, we were unable to find any diamonds in rock samples taken 2 centimeters below or above the Cretaceous-Tertiary Boundary or in samples taken 5 meters above or below. The diamonds are characteristic only of the Cretaceous-Tertiary Boundary rocks. In fact, we have looked for them near the Cretaceous-Tertiary Boundary rocks only of the Red Deer Valley in Alberta in part because these rocks were laid down on land, not under the sea. It is my opinion that the diamonds are so tiny that they would remain suspended indefinitely in sea water until destroyed by thermodynamical degradation and would never settle out to the bottom at all. I am not at all sanguine about the possibility of ever finding them in Cretaceous-Tertiary Boundary rocks laid down under the sea. Nevertheless, I regard their presence in the Cretaceous-Tertiary Boundary rocks of the Red Deer section as proof that the Cretaceous-Tertiary Boundary event involved a worldwide shower of diamonds, as well as of spinel of extraterrestrial origin.

Amino Acids

Amino acids are the universal building blocks of proteins, forming long chains. This is possible because one end of an amino acid is acid and the other is alkaline, so that the acid end of one can combine, forming a peptide bond, with the alkaline end of the next. Each such linkage ejects a water molecule from the peptide bond, and digestion of proteins largely involves splitting these peptide bonds by reinserting the water.

Two amino acids linked together by a peptide bond form a dipeptide. An example is the artificial sweetener aspartame, so beloved of dieters. A chain of a few amino acids is called an oligopeptide; an example is the hormone oxytocin, which induces the contractions of labor and childbirth. Add still more and we have a polypeptide, such as prolactin, which prompts the breasts to secrete milk. Chains of several hundred amino acids are called proteins.

The genetic code of DNA only codes for the inclusion of twenty amino acids in proteins, but the actual number found is about eight greater than this, because some amino acids are modified after the protein has been formed. For instance, an amino acid called hydroxyproline is common in structural proteins—that is, proteins found in the skin, the skeleton, and similar places. Hydroxyproline is not one of the DNA-coded amino acids but is formed from proline after this has been incorporated into a protein. It is also one of the flavor constituents in a good soup, made by boiling the structural proteins of bone and gristle to form a stock.

The acid part of an amino acid, at least in the amino acids that concern us here, is carboxylic, written -CO.OH, where C is carbon, O is oxygen, H hydrogen, and the dash at the beginning signifies attachment to the rest of the molecule (which can be written R- in the general case). The alkaline amino group consists of a nitrogen atom with two hydrogens attached: H_2N-. Thus, the general formula for any amino acid is H_2N-R-CO.OH. This is deceptively simple, for R can vary enormously and so can the precise point at which the amine group is attached. To make matters worse, amino acids can mostly exist in two forms, which are mirror images of each other, known as enantiomers. These are generally denoted by the prefixes D- and L- (from the Latin *dextro-* and *laevo-*), for right- and left-handed, respectively.

All protein amino acids are left-handed; acids synthesized in the laboratory by nonbiological processes are mixtures of left- and right-handed, a mixture known as a racemic mixture. If you were to leave a left-handed amino acid lying about for a few tens of thousands of years, the process known as racemization would eventually produce a racemic mixture, and indeed, this is the basis for one method of dating old bones—from the degree of racemization of the proteins in the bones. There are, however, just a few amino acids that will not spontaneously racemize in this way. One of them has been important to me in my research into the Cretaceous-Tertiary Boundary event, and I will return to it later.

Twenty-eight amino acids are found in our own bodies, 23 of them common to all animals and plants. All are left-handed, most are protein amino acids, several are breakdown products formed during metabolism, and one at least is a neural transmitter, serving to carry information from one brain cell to another. Some organisms have one or two amino acids not found elsewhere. An example is the toxic amino acids found in the nuts of cycad bushes. In addition, about 20 amino acids are unique to fungi, where they occur as essential components of antibiotics. In all, no more than 50 amino acids are known from living things; most of them are left-handed, though some fungal antibiotic amino acids are right-handed.

Carbonaceous meteorites may contain upward of 400 amino acids, far more than are known on earth—and this does not include left- and right-handed amino acids separately. If we did include them, the number would be about 800, since they are all present as racemic mixtures, except for a few that are symmetrical, neither left- nor right-handed. This is not the place to describe in detail how these amino acids are formed in outer space, but the raw materials—carbon, hydrogen, and oxygen—are abundant in comets, and given an energy source—a supernova say, or radioactivity—they could readily form. The precise pathways for their formation have been well studied.

Meteoritic amino acids are all relatively simple compounds; none of the more complex protein amino acids are found in meteorites. There are no "monounsaturated" or "polyunsaturated" amino acids in meteorites and no aromatic ones either, such as are found among the limited range of biological amino acids. Instead, they are all fully saturated—that is, they include the maximum number of hydrogen atoms the carbon atoms are capable of holding, and none of them contain an aromatic benzene ring such as we find in the common protein amino acid phenylalanine. Instead the carbon atoms in meteoritic amino acids are arranged in straight or branched chains, never in rings. A few do include two carboxylic groups instead of just one (a condition also found in some terrestrial amino acids), but otherwise there is no extra oxygen anywhere in any meteoritic acid. Thus serine, a common protein amino acid that contains an extra oxygen atom, is not found in meteorites; neither is hydroxyproline, the above-mentioned flavor ingredient in soups. (A soup made from carbonaceous meteorites would lack flavor.)

Amino acids have cumbersome names. A few of the more common acids have "trivial" names that say nothing about their structure. The structural names, awkward though they are, are generally more useful, since a scientist can envisage the structure just from the name. There are, in fact, two systems even for structural names. One of them calls the carbon atoms by Greek letters, starting with the one next to the carboxyl group. This is the more familiar way of naming, but it breaks down with some branched chains. The American Chemical Society has endorsed another method, that of numbering the carbon atoms, starting with the carbon of the carboxyl group as number 1. In this system the number 2 carbon is the α carbon of the Greek letter system. One might expect the alpha carbon to be the same as the number 1 carbon, but scientists are never as logical as that. In the Greek letter system of nomenclature, the carboxyl carbon (which is number 1 in the American system) is totally ignored.

The two most abundant fungal antibiotic amino acids are D-isovaline and AIB (for α-amino-isobutyric acid). The systematic names of these two acids are D-2-amino-2-methyl butanoic acid and 2-amino-2-methyl propanoic acid, respectively. Isovaline is one of those amino acids that cannot undergo spontaneous racemization, because of the presence of a methyl group (consisting of three hydrogen atoms attached to a carbon atom, written -CH_3) as well as an amino group on the number 2 carbon. AIB is symmetrical, not existing in separate left- and right-handed forms. These two compounds are also among the most abundant meteoritic amino acids, but isovaline is present in a racemic mixture, with left- and right-handed forms of the molecule in roughly equal abundance. Let me repeat: L-isovaline cannot be derived from D-isovaline by spontaneous racemization.

A few years ago I was investigating the part played by fungal antibiotics in inhibiting the biodegradation of waste from pulp mills by bacteria. Laboratory experiments suggested that bacterial degradation of the toxic wastes should proceed quite rapidly, but in the rivers it was occurring much more slowly. Perhaps the wood wastes were harboring colonies of fungi that would produce antibiotics and thus kill many of the bacteria. This indeed was what I found, and I analyzed the river water for the presence of the most common amino acids found exclusively in antibiotics. Not surprisingly, I found D-isovaline (but no L-isovaline) and AIB in abundance, even 20 kilometers downstream from a pulp mill. This was the first time these amino acids had been found in nature outside laboratory cultures of fungi, as I discovered when I searched the scientific literature. That same search also made me realize that these two amino acids were already well known from meteorites, together with L-isovaline. That was what started me on the research that has led to my writing this book.

Analytical methods for studying amino acids are among the most sensitive in common use, able not merely to distinguish but to measure concentrations as low as ten parts in a million billion (ten parts per quadrillion), to distinguish one amino acid from another at that concentration, and to tell left-handed from right-handed compounds. In addition, I found that these methods are easily automated, so that I was able to run the equipment overnight (in my spare time, since I was engaged in other research during the day) and pick up the results in the form of a computer disk in the morning. Automation was important for another reason: to reduce contamination. A single fingerprint or one flake of dandruff contains enough amino acids to ruin the analysis.

I used two main methods, both adapted from the research of numerous scientists before me and both depending on separation by chromatography. The first method is called high precision liquid chromatography (HPLC) and relies on separation of solutions of amino acids after reacting with a specific set of reagents; final detection depends on measuring the fluorescence given off under ultraviolet light. Figure 9.2 shows results of HPLC analysis of isovaline. The second method relies on separation as a gas of volatile derivatives of the amino acids and the measurement of the atomic masses of fragments of the molecules; this method is called gas chromatography–mass spectrometry (GCMS). The results of an analysis of isovaline by this method are shown in Figure 9.3.

Using these methods, Dennis Braman and I were able to identify and measure the concentrations of 51 amino acids in the Cretaceous-Tertiary Boundary rocks of Alberta. All of these were present as racemic mixtures, even those acids that cannot racemize spontaneously, and all had previously been recorded from carbonaceous meteorites. Eighteen of these amino acids were known only from these meteorites and have never been observed outside the laboratory before. Table 9.2 lists the amino acids that have been found only in carbonaceous meteorites and in the Cretaceous-Tertiary Boundary claystone. They were confined strictly to the Cretaceous-Tertiary Boundary, and were not present in rocks even 2 or 3 centimeters above or below the boundary. Figure 9.4 shows the concentration of one of these acids in a section across the boundary.

Traces of amino acids are common in sedimentary rocks, fossil chemical traces as it were, but these are only the more stable protein amino acids, those derived from dead animals, plants, and bacteria. Table 9.3 provides a classification of carboxylic amino acids, some of biological origin, others not. I have ignored the sulphuric amino acids (which have a sulphuric acid group instead of the carboxylic), some of which are of great commercial importance but which occur neither in organisms nor in meteorites. The rocks above and below the Cretaceous-Tertiary Boundary in Alberta are no exception. Just which amino acids are preserved in this way depends on the nature of the rocks. Glycine, the simplest of all possible amino acids, is ubiquitous in trace amounts, generally 2 or 3 picograms per gram, while in the Cretaceous-Tertiary Boundary layer its concentration may be a thousand times higher. Coal, in my experience, is marked by concentrations of aromatic and sulphur-containing amino acids, as

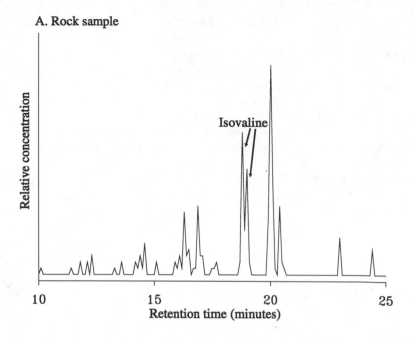

A. Rock sample

Relative concentration

Isovaline

Retention time (minutes)

10   15   20   25

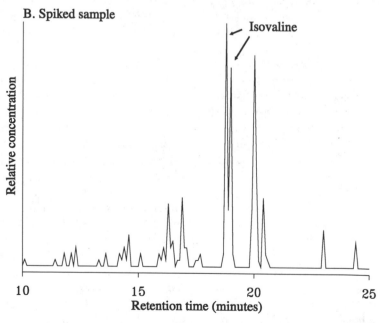

B. Spiked sample

Isovaline

Relative concentration

Retention time (minutes)

10   15   20   25

Fig. 9.2. High-precision liquid chromatography of isovaline. The result of analyzing a sample from the Cretaceous-Tertiary Boundary clay is shown in graph A; the
same sample to which a small amount of genuine (synthetic) isovaline had been
added as a so-called spike is shown in graph B. Note that both right-handed and
left-handed forms of this amino acid are present.

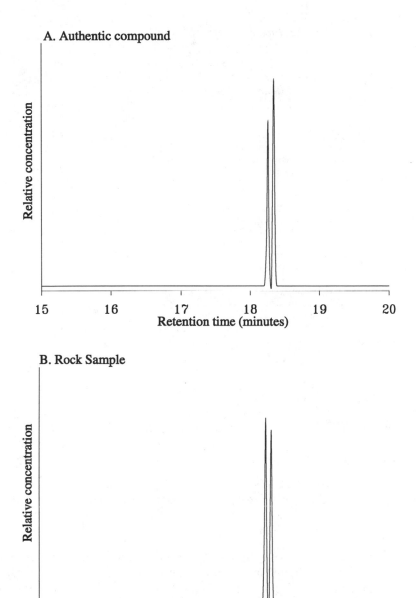

Fig. 9.3. Gas chromatography–mass spectrometry of isovaline. The upper trace was obtained from a synthetic mixture of the right- and left-handed forms of this amino acid; the lower, from a sample prepared from Cretaceous-Tertiary Boundary claystone.

TABLE 9.2
Amino Acids Found in the Cretaceous-Tertiary
Boundary Claystone of Alberta

Acid	Group[a]
2-amino pentanoic acid (norvaline)	1.1
3-amino pentanoic acid	1.1
4-amino pentanoic acid	1.1
L-2-amino-2-methyl butanoic acid (L-isovaline)[b]	1.2
3-amino-2-methyl butanoic acid	1.2
4-amino-2-methyl butanoic acid	1.2
3-amino-3-methyl butanoic acid	1.2
4-amino-3-methyl butanoic acid	1.2
3-amino-2-ethyl propanoic acid	1.2
3-amino-2,2-dimethyl propanoic acid	1.2
2-amino-2-methyl pentanoic acid	1.2
1-amino-cyclopentane-1-carboxylic acid (cycloleucine)	1.3
N-methyl-2-amino propanoic acid (N-methyl alanine)	1.4
N-ethyl-2-amino ethanoic acid (N-ethyl glycine)	1.4
2-methyl-2-amino butandioic acid (2-Me aspartic acid)	4
3-methyl-2-amino butandioic acid (3-Me aspartic acid)	4
2-Me-2-amino-1,5 pentandioic acid (2-Me glutamic acid)	4
2-amino-1,7 heptandioic acid (2-amino pimelic acid)	4

NOTE: All of these acids, which have no other known terrestrial occurrence, exist in racemic proportions in the rocks and in concentrations (relative to iridium) roughly equal to that found in the Murchison meteorite.
[a]See Table 9.3.
[b]D-isovaline is known from fungal antibiotics but cannot give rise to the laevo form by spontaneous racemization.

shown in the following list of the amino acids found in the coal seams immediately overlying the Cretaceous-Tertiary Boundary but which are absent from the boundary claystone itself:

Aromatic acids:	Sulphur-containing acids:
Phenylalanine	Cysteine
Tyrosine	Methionine
Tryptophan	

Neither of these types of amino acid is found in the Cretaceous-Tertiary Boundary layer itself, just in the coal above.

Instead, the Cretaceous-Tertiary Boundary rocks contain those amino acids characteristic of carbonaceous meteorites and, with one exception, in the same proportions as in the best-studied such meteorite. The exception is protein amino acids, which occur in much lower concentrations than one might predict from the concentrations in meteorites. Table 9.4 lists some nonprotein amino acids found in the boundary rocks and in meteorites.

This is easily understood when we attempt to culture bacteria, offering them only selected amino acids as an energy source: the bacteria flourish

TABLE 9.3
A Functional Classification of Carboxylic Amino Acids

Group	Examples
1. Mono amino alkanoic acids	
1.1 Linear	Alanine
1.2 Branched	Valine
1.3 Cyclic	Proline
1.4 N-alkyl	MeDAP[a]
2. Mono amino alkandioic acids	Aspartic acid
3. Mono amino aromatic acids	Phenylalanine
4. Poly amino acids	Arginine
5. Substituted acids	
5.1 Carbamyl	Asparagine
5.2 Hydroxyl	Hydroxyproline; serine
5.3 Sulphur-containing	Cysteine
5.4 Iodinated	Thyroxine

[a] This is the only group in this classification that contains no protein amino acids; the example given is one of the neurotoxins found in cycad nuts.

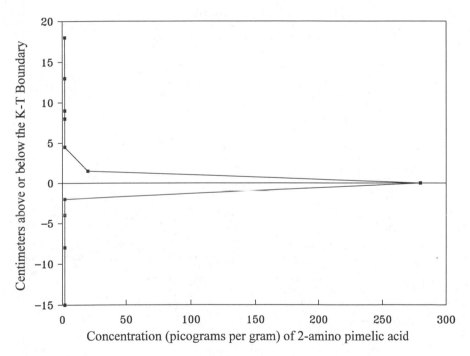

Fig. 9.4. Concentration of 2-amino pimelic acid near the Cretaceous-Tertiary Boundary.

TABLE 9.4

Some Nonprotein Amino Acids Found in the Cretaceous-Tertiary Boundary of Alberta and in Meteorites

Acid	Boundary to meteorite ratio
2-amino pentanoic acid	1.10
3-amino pentanoic acid	1.04
4-amino pentanoic acid	0.98
L-2-amino-2-methyl butanoic acid	0.93
3-amino-2-methyl butanoic acid	1.06
4-amino-2-methyl butanoic acid	0.95
3-amino-3-methyl butanoic acid	1.11
4-amino-3-methyl butanoic acid	0.90
3-amino-2-ethyl propanoic acid	0.99
3-amino-2,2-dimethyl propanoic acid	1.00
2-amino-2-methyl pentanoic acid	1.02
1-amino-cyclopentane-1-carboxylic acid	0.96
N-methyl-2-amino propanoic acid	1.01
N-ethyl-2-amino ethanoic acid	1.00
2-methyl-2-amino butandioic acid	0.92
3-methyl-2-amino butandioic acid	0.89
2-Me-2-amino-1,5 pentandioic acid	1.07
2-amino-1,7-heptandioic acid	1.10

TABLE 9.5

Protein Amino Acids in the Cretaceous-Tertiary Boundary of Alberta and in Meteorites

Amino acid	I[a]	II[b]	III[c]	IV[d]	V[e]
Alanine	0.11	0.04	41	7.01	0.01
Aspartic acid	0.13	0.05	5	0.85	0.06
Glutamic acid	1.32	0.53	68	3.08	0.17
Glycine	0.10	0.04	98	16.75	0.01
Isoleucine	0.21	0.08	4	0.68	0.12
Leucine	0.19	0.08	4	0.68	0.11
Norleucine	0.11	0.04	2	0.34	0.13
Proline	0.24	0.10	16	2.74	0.04
Valine	0.31	0.12	8	1.37	0.09

[a] Nanograms per square centimeter (combined enantiomers) in Cretaceous-Tertiary Boundary rocks, integrated over the rock column.

[b] Ratio of amino acid to iridium in rock, integrated over the rock column.

[c] Nanomoles per gram in the Murchison meteorite.

[d] Ratio of amino acid to iridium in the meteorite.

[e] The ratio of ratios (i.e., column III divided by column V). Relative to iridium, the rocks have far less of these protein amino acids than does the meteorite. (This ratio would be equality if they had the same relative amounts.)

on a diet of protein amino acids but do not if they are offered the exotic forms. I conclude that any protein amino acids originally in the rocks have been partly digested away by bacterial action, resulting in a lower-than-expected concentration in the rocks today. Table 9.5 lists protein amino acids found in boundary rocks and in meteorites.

If we compare the concentration of the exotic amino acids with the iridium level in the rocks, we find that this too agrees with the same ratio, taken acid by acid, found in the Murchison meteorite, the best-studied of the carbonaceous meteorites. The same applies to the levels relative to the nanometer-sized diamonds.

Finally, the amino acids extracted from the Cretaceous-Tertiary Boundary rocks have a carbon isotope ratio similar to that of the diamonds and to that of interstellar dust and are quite different from the amino acids extracted from the overlying coal deposits, where the carbon isotope ratio is that of terrestrial material. In other words, both the amino acids and the diamonds have come from outer space.

The only possible conclusion is that the iridium and the amino acids as well as the diamonds were delivered to earth by the impact of a bolide whose composition was roughly the same as that of the Murchison meteorite.

The Answer . . .
and More Questions

From the evidence presented in this chapter, much of it the result of my own research, it is clear that the Cretaceous-Tertiary Boundary event could not have been brought about by the impact of an asteroid or asteroid fragment, whether a stony or a metallic meteorite in nature. The carbon and carbon compounds found in the Cretaceous-Tertiary Boundary rocks could only have come from a cometary impact. Diamonds and amino acids are not components of asteroidal material. Instead they are found in abundance in comets and in their derivatives the carbonaceous meteorites. The delivery vehicle that introduced them to the earth at the time of the Cretaceous-Tertiary Boundary event can only have been cometary in nature, especially since the carbon isotope ratios of these materials are all those of interstellar objects, not those of the earth and of the asteroids. In addition, only a comet would break apart quickly enough on impact that its interior would not reach temperatures high enough to destroy both diamonds and amino acids. Asteroid fragments would hold together far longer and the heat on impact would be much higher. Figure 9.5 shows calculations for the temperature reached by a bolide when hitting the earth. The temperature reached by the bolide depends on its tensile strength, while the temperature reached in the bedrock depends on the strength of the bedrock. In impact craters the strength of the rocks usually limits the temperature at the bottom of the crater to around 3,500 to 4,000 Kelvin, and it has often been assumed that this is also true of the

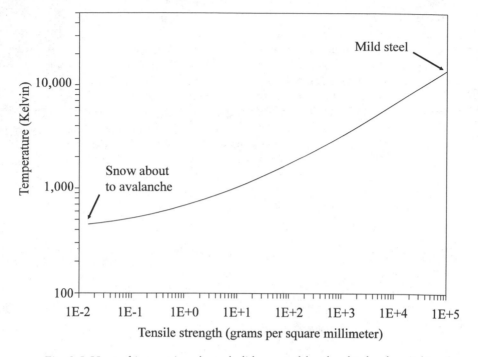

Fig. 9.5. Heat of impact in a large bolide created by the shock when it hits the earth. The temperature reached by the bolide depends on its tensile strength, which would probably fall between that of a giant snowball and that of mild steel, depending on the composition of the bolide. This figure shows the calculations made for a bolide of 1 percent of the mass of the Cretaceous-Tertiary Boundary impactor hitting the earth at a velocity of 75 kilometers per second.

temperature reached by the material of an incoming bolide before it is blown apart. This is true if the bolide is made of something similar to earthly sandstone (such as a stony meteorite), but if it is made of mild steel (which approximates the strength of an iron meteorite), then its interior can reach as high as 10,000 to 14,000 Kelvin. At the other extreme, a giant snowball would self-destruct at a temperature no higher than a few hundred Kelvin (no hotter than a kitchen oven).

This still leaves two major questions open. Was the impactor a "live comet" fresh in from the Oort Cloud, still fizzing like a bottle of champagne? Or was it a "dead comet," one from which the volatile materials had been baked off by the prolonged heat of the sun as it was trapped in the asteroid belt as a secondary carbonaceous asteroid? And was the Cretaceous-Tertiary Boundary event a single impact from a single such bolide, or was it, as Piet Hut has suggested, a veritable shower of comets? The answers to these questions require a consideration of the energetics of comets, which is the subject of the next chapter.

10

The Energetics of Impactors

Think how once, when Yao was heaven,
Ten suns burnt the nine regions,
Metals gushed into streaming quicksilver,
Jade cooked to charred cinnabar,
The baking universe became a kiln. . . .
And in sudden fury let loose a mighty torrent.
—Lu T'ung, *Poems of the Late T'ang*
(translated by A. C. Graham)

The majority of meteors and meteorites enter the Earth's atmosphere at a relative (geocentric) velocity of about 18 kilometers per second. This is about 650 times the velocity of a car on the highway, or about 75 times the speed of a Boeing 747 at cruising altitude.

It is one of the paradoxes of celestial mechanics that the slower an approaching object is moving, the less is its chance of actually colliding with the earth. In fact if its approach velocity is less than 11.2 kilometers per second such an object cannot even hit the Earth or enter the atmosphere; it will simply be shouldered aside and its trajectory altered. An incoming space shuttle above drag of the atmosphere actually fires its rockets to accelerate in order to reenter the earth's atmosphere (not retrorockets, beloved of science fiction, for the shuttle has none). Once in the atmosphere, aerodynamic drag will slow it down, but by then the shuttle is in the grip of the mother planet.

Escape Velocity and Earth-Impacting Objects

This magic figure of 11.2 kilometers per second is, of course, the escape velocity from the earth, the velocity any rocket (or any other object for that matter) must achieve in order to escape from earth's gravity. It is a barrier that prevents any outgoing object from getting away if it is moving too slowly, but also one that prevents any slow-moving object from coming in.

The escape velocity depends on the mass of the planet. The escape velocity for a lunar landing module to rise from the surface of the moon is far less than the terrestrial escape velocity, since the mass of the moon is so much lower—only one-eightieth that of the earth—and its escape velocity is barely one-fifth of that of the earth, some 2.37 kilometers per second. Conversely, Jupiter's escape velocity is much higher than the earth's, 60.24 kilometers per second, or over five times that of the Earth, since Jupiter's mass is about 316 times that of the earth. One consequence of this is that over its history the moon has been hit (and hence cratered) far more frequently by asteroids and other objects than has the earth, while Jupiter has rarely if ever been hit. In fact, comets passing near Jupiter in free-fall can never hit that planet but are deflected and their orbits tilted. The fragmented object that hit Jupiter in July 1994 was not in free-fall around the sun, but rather in orbit around Jupiter. Technically speaking, it was a moonlet of Jupiter, certainly not a comet. It was most probably of asteroidal origin, and the apparent coma that made it appear something like a comet when observed telescopically seems to have been dust generated when the object was broken apart on its last pass around Jupiter.

Consider for a moment a comet coming in toward the earth from the Oort Cloud without passing anywhere near Jupiter (or any of the other gas giants). Since it must have left the Oort Cloud roughly in the plane of the ecliptic, it will approach the orbit of the earth still in that plane, not tilted significantly in any way. Moreover, it will be orbiting the sun in the same sense as the earth—counterclockwise, as seen from the north pole of the solar system. At the point where it crosses the earth's orbit, such a comet in free-fall, moving under gravity alone, will have a heliocentric velocity (meaning a velocity relative to the sun) of about 25 kilometers per second; the precise figure depends on the eccentricity of its orbit. But the earth, going around the sun in the same direction, though in a much more nearly circular orbit, will have a heliocentric velocity of about 29 kilometers per second (give or take a few hundred meters or so, depending on the precise point in its orbit). The approach velocity of the comet to the earth—that is, the geocentric velocity (or velocity relative to the earth)—will be the difference between these two, about 4 kilometers per second, since they are moving in roughly the same plane and in the same direction.

Obviously, a virgin comet—one in fresh from the Oort Cloud, untilted by an approach to Jupiter and moving in free-fall—cannot possibly hit the earth. Its approach is too slow. The actual result of what appears to be a collision course will be that the comet is shouldered aside, its trajectory altered, and its orbit tilted.

Consider now a comet that has passed so close to Jupiter on its way in toward the center of the solar system that its orbit has been flipped completely over through 180 degrees so that it is in what is called a retrograde orbit. It is now moving in the plane of the ecliptic but orbiting the sun in the opposite direction from the earth. Its velocity is little changed and it crosses the earth's orbit at the same heliocentric velocity of 25 kilometers per second. The precise figure depends on the nature of the orbit, primarily on its degree of eccentricity—in other words, on how far from a circle the orbit departs, or how oval the orbit is. If such a comet were on a collision course with the earth, the comet would be coming head on. The geocentric velocity of this is the sum of the two heliocentric velocities, not the difference. Impact would take place, and the impact velocity would be the maximum possible for an object in free-fall, or some 56 kilometers per second.

When we consider intermediate conditions, with the orbit of the incoming comet tilted by varying amounts, we find that impact cannot occur at all (if the bolide is in free-fall from the Oort Cloud) unless the angle of its orbit is greater than 46 degrees to the plane of the ecliptic, which can only happen if its earlier trajectory has taken it quite close to Jupiter, as the graph in Figure 10.1 shows.

There is one other possible way in which Jupiter can alter the trajectory of a passing object. If it passes close in front of Jupiter (in a prograde, or direct, orbit), it will lose enough orbital energy so that it is captured into a short-period orbit (an orbit that takes, by definition, less than 200 years) with higher binding energy to the sun. On the other hand, if it passes close behind Jupiter the comet can take orbital energy from the rotation of Jupiter and be accelerated as well as tilted in its orbit. This is known as the slingshot effect. Its overall result is to sling the comet on a hyperbolic orbit away from the sun. It is almost impossible that its new orbit will take it anywhere within the zone of the stony planets. NASA has taken advantage of this slingshot effect on a number of occasions. Voyager 2, for instance, received an orbital energy boost from a backside passage of Jupiter to enable it to reach Saturn and Uranus successively. Backside passages of these two planets further boosted the orbital energy of the probe until now it is on a hyperbolic orbit that will take it right outside the solar system, passing through the Oort Cloud in many scores of thousands of years, never again to return to the solar system.

Combining now the statistics of the relationship of the orbits of comets with respect to Jupiter with the angle of tilt that this will induce, we find that the peak frequency of impacts of comets with the earth occurs at an

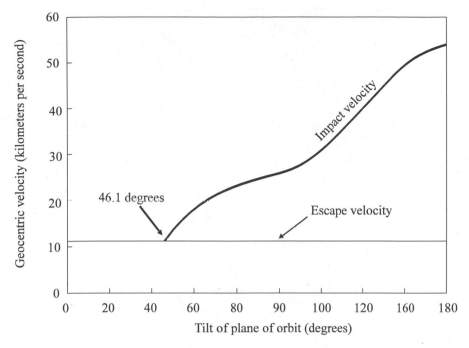

Fig. 10.1. Impact velocity of a bolide relative to the tilt of the plane of its orbit.

angle of approach of about 60 degrees to the plane of the ecliptic. As is illustrated in Figure 10.2, the largest number of bolides hit at around 16 kilometers a second, and the mean velocity of impact is just over 18; the maximum is about 55 kilometers per second. These are theoretical calculations based on the effects of tilting the orbits of comets.

A comet in free-fall, then, can only hit the earth if (1) its trajectory has been tilted by close approach to Jupiter or to one of the other gas giants, (2) its orbit is tilted by more than 46 degrees, and (3) its approach velocity is greater than 11.2 kilometers per second.

It should be noted that an asteroidal meteorite may not be in free-fall like a comet, at least on its first orbit of the sun, since the violent birth of such a meteorite in the disruption of an asteroid may impart an initial higher velocity to its trajectory. Accordingly, it may cross the earth's orbit at a higher velocity than the 25.1 kilometers per second dictated by gravity. This does not, however, alter the 11.2 kilometers per second lower limit set upon impact velocity by the escape velocity of the earth.

This all sounds terribly abstruse, but it is relevant to the current watch, now being established, for possible earth-impacting objects circling the sun. About 2,000 objects with earth-crossing orbits have been identified

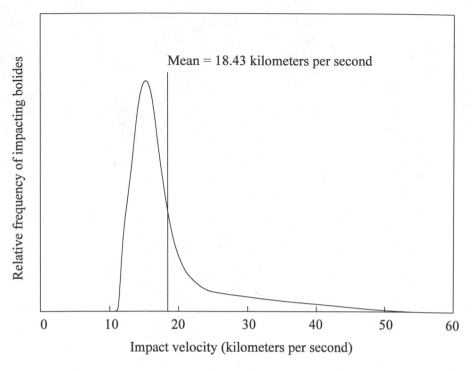

Fig. 10.2. Impact velocity of a bolide in free-fall.

larger than about 1.5 kilometers across; the largest are the Apollo bodies, or dead comets. It would be an impossible task to monitor the orbits of each and every one of these objects, let alone the even more numerous smaller objects, but there is no need to do so, since such an object cannot hit the earth unless its orbit is tilted far out of the plane of the ecliptic. Only these objects—a small percentage of the total—need to be monitored if we are to have early warning of their approach and possibly devise some means (perhaps thermonuclear warheads on rockets) to deflect them from a disastrous collision course. Who says theoretical astronomy is a useless pursuit?

Speed of the Bolide

How fast was the Cretaceous-Tertiary Boundary bolide moving? This question is meaningless unless we specify the frame of reference we are using to measure the velocity. It should be obvious that we need the answer expressed in geocentric terms: What was the approach velocity of the bolide, the comet, in earth-centered terms? We can begin to obtain some kind of answer to this question by considering what evidence we have for the mass of the bolide and for the mass of the Cretaceous-Tertiary Boundary layer in the rocks.

From the moment when the impact hypothesis was first proposed by the Alvarez group at the University of California at Berkeley, it has been clear that we are dealing with an impactor of some 10^{18} grams, or a thousand billion metric tons. We shall see later that this is an oversimplification. As I have argued above, this would correspond to a chunk of iron about 6.5 kilometers across, a chunk of rock about 10 kilometers across, or to a comet roughly 25 kilometers in diameter (a comet is much less dense than a stony object).

This figure of 10^{18} grams was first calculated from the amount of iridium found in the Cretaceous-Tertiary Boundary rocks. Indeed, it was the anomalously high levels of this element that led first to the hypothesis that the Cretaceous-Tertiary Boundary layer represented ejecta from the impact of an asteroid. As described in Chapter 9 this calculation has been repeatedly refined as other exposures of the Cretaceous-Tertiary Boundary rocks have been studied by many teams of scientists, and the conclusion remains that the mass of the impacting bolide was indeed between 10^{18} grams and 2×10^{18} grams.

More recently the discovery of other measurable cometary debris in the Cretaceous-Tertiary Boundary layer has confirmed this figure. Calculations from the concentration of tiny diamonds and from the levels of amino acids all point to a bolide of mass 10^{18} grams or greater, in particular, as shown in the Chapter 9, they point to cometary material of about 2×10^{18} grams.

The slight indeterminacy found here stems from the composition of the bolide. The original calculations were made from the composition of metallic meteorites, which are enriched twofold in iridium compared with the primitive composition of the solar system. The figure of 10^{18} grams is only correct if the impacting bolide was indeed a metallic meteorite. In that case the heat generated would be just as imagined by Lu T'ung, the ninth-century T'ang dynasty poet, when the ten suns burned the nine regions of China and metals (including iridium) fused to streaming quicksilver. If instead it was, as I have argued, a comet, whose content of iridium is the primordial level for the whole solar system, then the total mass must have been twice this value to provide the same amount of iridium, and the heat of impact would have been much lower.

The next stage in the calculations is to determine just how much material was thrown up as a stratospheric cloud. This material included the dust produced from the disintegration of the bolide itself as well as rock dust excavated in the creation of the impact crater. The immediate rim of the crater itself and the tektites thrown out into the near vicinity ("near" means within a few hundred kilometers) are excluded from the first stages of this calculation because this material never reached the stratosphere. The great cloud of dust that later settled down on the earth, however, to create the worldwide Cretaceous-Tertiary Boundary layer can only have traveled around the entire globe if it had penetrated into the stratosphere. If it had remained lower in the atmosphere (as did the smoke cloud from the burning oil wells of Kuwait), it could never have spread around the world.

Except in the vicinity of the great crater in Yucatán, where the Cretaceous-Tertiary Boundary layer is thickened to hundreds of meters by tektites and rock flows—material that never reached the stratosphere—the layer varies from about 8 millimeters to as much as 30 centimeters thick, with a mean of 2.2 centimeters. A worldwide layer of rock 2.2 centimeters thick represents about 2.5×10^{19} grams of material, either 12 or 25 times the mass of the impacting bolide, depending on what you believe the mass of that bolide to have been. How much energy is needed to loft this much material to the stratosphere? This is a straightforward sum, one which you can almost do in your head. The answer is about 1.5×10^{26} joules.

Now all this energy must come from the energy of impact of the bolide—that is, the comet. In addition, this impact must release enough energy to account for the heat developed on impact, heat great enough to melt rocks, form tektites, excavate the crater, and, finally, throw the ejecta blanket and the tektites up into the lower atmosphere. Nevertheless, much of the energy from the impact is dissipated in throwing dust up to the stratosphere. If at first we ignore the remaining energy we shall not go far wrong; in any case we know that the calculation made using this simplification will supply a minimum value for the energy of the impact.

The impact, then, released something over 1.5×10^{26} joules of energy—actually something over this figure, because this amount of energy was used to impel the dust into the stratosphere and does not take into account the heat generated and the energy that went into forming the ejecta blanket. A mass of 2×10^{18} grams must have been traveling at 78 kilometers per second (relative to the earth) to release this much energy on impact. If now we take into consideration the other energy we have so far neglected, then the energy released was over 2×10^{26} joules, and the comet must have been traveling (in geocentric terms) at considerably over 80 kilometers per second. This is true whether we are dealing with a single mass of 10^{18} grams or with this total mass divided among several objects. In the latter situation, the mean velocity of all the impacts exceeded

80 kilometers per second. Thus, the bolide, or bolides, that wreaked the havoc at the Cretaceous-Tertiary Boundary hit the earth with a speed greater than 80 kilometers per second.

The Accelerating Mechanism

This impact velocity of 80 kilometers per second is impossible for an object in free-fall. Therefore, the conclusion is inescapable that the comet (or comets) whose impact left the debris that forms the Cretaceous-Tertiary Boundary layer was traveling at a speed that cannot have been the result of gravity alone. This was not merely a perturbed object left to fall freely toward the sun but an object that had been strongly accelerated.

There are only two plausible mechanisms for such an acceleration, and one of these applies only to asteroids. The tidal forces caused by Jupiter and the galactic center (the mass of stars and possibly a black hole at the center of our galaxy that exerts a significant gravitational pull on the whole solar system) disturb the orbits of asteroids so that they occasionally collide with each other. A gentle collision may result in fusion between two asteroids, but a violent collision may lead to disruption of the colliding bodies, resulting in smaller pieces being flung off at high speeds. In this way an asteroid fragment could cross the orbit of the earth at a velocity significantly higher than free-fall. Indeed, some meteorites of asteroidal origin have been observed to be traveling at such velocities.

This mechanism, however, is only applicable to asteroids, not to comets, and only relatively small pieces of asteroids could reach the velocities we have discussed above—pieces smaller than about 1.5 to 2 kilometers across. The bolide that caused the Cretaceous-Tertiary Boundary catastrophe could not have been accelerated by this mechanism. Instead, only a nearby supernova could have provided the momentum and energy necessary to accelerate a comet to these velocities. If we accept that the destruction wrought at the Cretaceous-Tertiary Boundary event was the result of cometary impact, then the conclusion seems inevitable that in some way a supernova was involved.

The Alvarez team, in their seminal 1980 paper, considered the possibility that the iridium anomaly resulted from a supernova but dismissed the idea because such an event would have to have been too close. For a supernova to have been close enough to provide all the iridium found at the Cretaceous-Tertiary Boundary, it would have destroyed the whole solar system or at least stripped the planets from the sun and thrown them into outer space.

I am not suggesting that a supernova occurred as close as that. On that scale of distance, measurable in milliparsecs, the supernova I am suggesting was relatively distant, not within the Oort Cloud as a closer supernova would have been, but well outside, a little farther away than is the nearest star today. It would have been too far away to have contributed a measurable amount of iridium to the Cretaceous-Tertiary Boundary rocks (measurable, that is, over and above what must have been contributed by cometary impact) but near enough to have imparted a good extra thrust to comets in the Oort Cloud, to have transferred momentum to them and accelerated them well above free-fall velocity.

No, the bulk of the Cretaceous-Tertiary Boundary iridium anomaly came from the bolide, as the Alvarez team suggested, but in addition the presence of the supernova may be detectable in other isotopic anomalies, as we shall see.

Supernovae and
Cometary Acceleration

And behold a great red dragon, having seven heads and ten
horns and seven crowns upon his heads. And his tail drew the
third part of the stars of heaven and did cast them to the earth.
—Revelation 12: 3–4

The stars of heaven cast down to the
earth in this way can only have been a meteor shower, like the summer
Perseids or the November Leonids.

One of the earliest records of a meteor shower is found in the Chinese
classic *Ch'un Ch'iu* (Spring and autumn annals) sometimes ascribed to
Confucius (or Kung Fu-tzu), which describes a meteor shower in 687 B.C.
Some meteor showers, such as the Perseids, occur regularly every year on
the same date. One of the longest-known showers—known for more than
a thousand years—is the Leonids, which appear every year to radiate out
from a point in the constellation Leo. Most years they are quite faint, but
every 33 or 34 years they reach a peak in numbers. In the Orient these
peak displays have been observed three times a century since 902 A.D.

A particularly brilliant display by the Leonids in 1833 was described
by a South Carolina plantation owner:

I then opened the door, and it is difficult to say which excited me most—the aw-
fulness of the scene, or the distressed cries of the negroes. Upward of one hundred
lay prostrate on the ground—some speechless, and some with the bitterest cries,
but with their hands raised, imploring God to save the world and them. The scene
was truly awful; for never did rain fall much thicker than the meteors fell towards
the Earth.

I observed the Leonids myself in the Butana Desert of northern Sudan on November 17, 1966, one of the peak years of the 33-year cycle. This desert is too far south to view the meteors at all well, for Leo is a northern constellation, but I estimated that at the peak (at about two o'clock in the morning) there were about 40 meteors per second; the same estimate was made that year in Arizona. Three weeks later I met, for the first time, my wife, Roxane, an ethnomusicologist, a chance encounter in the Butana where we were both carrying out field studies.

Sixty-six years earlier the Leonids had caused a panic in London. Observers claimed to be able to read a newspaper by the light of the meteors; the less educated feared the end of the world. Riots broke out and several people were killed.

Showers of Meteors and Comets

A meteor shower marks the passage of the earth through the trail of dust left behind by the passage of a comet in its orbit around the sun. The meteors are mere particles of dust, most less than a millimeter across, some as small as a micrometer, leaving visible trails at elevations between around 120 and 90 kilometers high in the atmosphere as they burn up and ionize the air around them. It is extraordinary that such small particles can leave trails visible at a distance of 100 to 150 kilometers; the fact that they do gives some indication of the speed and hence the energy with which they arrive. It is doubtful if any fragment of cometary dust contributing to a meteor shower ever penetrates to within 20 kilometers of the earth; most of them burn up at over 90 kilometers.

A recent analysis of the Tunguska event, which occurred in Siberia in 1908, has led to the conclusion that any cometary object less than a kilometer across could never penetrate to the surface of the earth; it would blow apart before reaching an altitude of about 80 kilometers. The authors conclude that the object that devastated Siberia was more likely to be a fragile stony bolide, a fragment of asteroidal crust about 100 to 150 meters across, which blew apart at an altitude of about 10 kilometers. Since it did not reach the surface of the earth, it was not by definition a meteorite but a fireball.

Potentially more damaging than meteor showers are meteorite showers, since by definition meteorites are bolides that actually reach the surface of the earth, much slowed down by their passage through the atmosphere but not completely burnt up. These probably arise from the breakup of a large fragile bolide (probably a fragment of a comet or of a meteorite) high in the atmosphere. Holbrook, Arizona, witnessed a meteorite shower in 1912, when an estimated 14,000 meteorites fell; in 1868 about 100,000 fell in the neighborhood of Pułtusk, Poland, without doing significant damage.

An intact comet, however, or a shower of comets, "cast down to earth" at these same initial velocities can do enormous damage. A large bolide is decelerated hardly at all by passage through the atmosphere. But what "great red dragon" could cast down a comet—or a shower of comets—to the earth?

Theories of Cometary Perturbation

I have argued earlier that the comet (or comets) whose collision with the earth at the time of the Cretaceous-Tertiary Boundary event was moving too fast for free-fall—that is, it could not have moved that fast under the influence of gravity alone. Something had to give it a good kick in the pants and accelerate it to higher speeds. Some great red dragon had breathed fire on it and booted it toward the sun and the earth. What are the various hypotheses that have been advanced to account for comets being thrown out of their orbits in the Oort Cloud?

Passing Stars

First, there is the idea that a star passing close enough to the sun could perturb the orbit of comets, which has certainly happened repeatedly in the past. Figure 11.1 is a plot of the theoretically calculated average time intervals between the closest approaches of passing stars. It is important to observe that these approaches occur at irregular intervals, not periodically. Although on average a star comes within the inner limit of the Oort Cloud a little more often than once every 30 million years, this does not happen periodically, but rather stochastically—in other words, it happens at chance intervals.

When a star passes through the Oort Cloud, or even comes near the outer limit of it, the orbits of all the comets are perturbed so that they begin to circulate not around the sun but around the joint center of mass of the sun and the star. The position of this joint center is continually changing as the stars pass each other, making for rather weird gyrations on the part of the captive comets. As the two stars (or the sun and the star if you prefer) draw apart again, some of the comets remain with the visiting star. None of them is perturbed in such a way that they move toward the inner solar system. The net effect is that the passing star has stolen some of the sun's Oort Cloud.

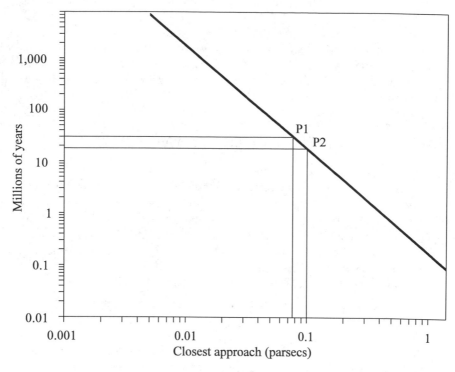

Fig. 11.1. Frequency of closest approach of passing stars, a theoretical calculation. The point P1 marks the 26-million-year interval postulated by believers in periodicity, a distance of about 0.073 parsecs, or 73 milliparsecs; P2 marks the frequency with which stars pass within the inner edge of the Oort Cloud (0.1 parsecs), roughly every 15 million years.

Stars have passed close enough for this to happen sufficiently often for the whole Oort Cloud to have been lost in the last four and a half billion years, if this were the only process taking place. The fact that this has not happened and that the Oort Cloud still contains hundreds of billions of comets indicates that the cloud has been replenished. The only way this could happen is if every star in the local cluster, at least every star of the same spectral class as the sun, has its own Oort Cloud. Indeed, I have shown in Chapter 5 that the nature of the formation of the solar system implies that all stars in the local cluster formed in the same way as the sun must have such a cloud, whether or not they also have planetary systems too.

The close passage of another star ("close" in this context means near enough to perturb the orbits of comets in the Oort Cloud) results in an exchange of comets between the clouds of the sun and of the passing star. It will not result in perturbation of comets into the inner solar system. This mechanism cannot impart a high velocity to a comet.

Indeed, some such mechanism could well be the reason why comets do not all have the same composition or exactly the same age: they may have originated around different stars. Moreover, the present-day Oort Cloud contains far too many comets for them all to have originated around our sun; far more must have been lost to passing stars. Accordingly, rather than losing comets to every passing star, the sun may simply have traded a few each time—a sort of cosmological swap meet of stars. Recently I have shown that the Allende meteorite, which is a fragment of a comet, is about 70 million years older than the solar system, and I conclude that it must have formed around another star.

In the remote event that a comet is perturbed by this kind of close passage of another star into a free-fall trajectory that will take it into the inner solar system, it will take about 330,000 years to get to earth. After all, the orbital period of a comet in the Oort Cloud is about a million years, and there is only gravity to accelerate it. Moving at this slow speed and slow rate of acceleration, the comet is most likely to be trapped in the Kuiper Belt, the parking orbit outside Neptune, having been captured by the gravitational field of the gas giants. It could even enter the asteroid belt and become a carbonaceous asteroid trapped by the gravity of Jupiter. This is probably what happened to the parent body of the Allende meteorite, which landed in Mexico in 1969.

In the extremely remote event that Jupiter diverted such a comet into an earth-crossing trajectory, the chances of an actual collision are remote. For one thing, such a comet, diverted by Jupiter, would be in a highly inclined orbit and is likely to cross the plane of the ecliptic (the plane in which the orbit of the earth lies) at some point (or rather at a pair of points, one on its way in and one on its way out) that does not coincide exactly with the orbit of the earth. In other words, it would pass the earth's orbit either above or below. For another, if it were on a collision course with the earth, it is still unlikely that an actual collision would take place; the comet is moving too slowly, relative to the earth.

For a NASA space probe to leave the earth, it must attain terrestrial escape velocity, approximately 11.2 kilometers per second. Below that velocity it will fall back to earth again without ever reaching outer space. The converse is also true: a body must be moving (relative to the earth) at a speed above the escape velocity to come down to earth.

A comet approaching the earth at less than the escape velocity will simply be shouldered out of the way. That is why few comets ever hit

Jupiter. Jupiter's mass (316 times that of the earth) and therefore its escape velocity (about 60 kilometers per second) is higher than any speed a comet in free-fall could reach. All comets approaching Jupiter are simply perturbed into new highly tilted orbits. It is much easier for a comet to hit the moon, with its much lower escape velocity. That is one reason why we cannot extrapolate from cratering on the moon to possible early cratering on the earth.

Most free-fall comets will be moving too slowly (with respect to the earth, which is itself hurtling around the sun) ever to hit the earth, even if they are on a collision course. They will just be shouldered out of the way, diverted, deflected, their orbits tilted—and they may even be ejected entirely from the solar system in a hyperbolic orbit.

Nemesis

A second hypothesis has been named the Nemesis hypothesis. Unlike the close approaches of other stars, approaches that are known to happen, this theory has been advanced solely in order to explain the perturbation of the Oort Cloud and has no other raison d'être. It was in fact advanced in order to account for the supposed periodicity of mass extinctions, a periodicity we now know to be spurious, a result of the poor use of statistics. The old saw is right in this case: there are lies, damned lies, and statistics.

According to the Nemesis hypothesis the sun is not alone, but has a companion star; in other words, the sun is one member of a binary pair, like so many stars visible with a telescope. Its twin star, Nemesis, is supposed to be much smaller than the sun and to be orbiting it in a very eccentric orbit. In fact, both stars would be orbiting their common center of mass, but Nemesis is supposed to be so much smaller than the sun that this is a convenient shorthand way of stating it. The orbit of Nemesis is supposed to be so eccentric that it approaches the inner solar system every 26 million years (the interval proposed between successive mass extinctions) and so perturbs comets into earth-crossing orbits at this interval.

It is an easy exercise to compute the mass and orbital parameters of Nemesis that would cause it to behave like this. At the present day, such a star would be about 4 parsecs (13 light-years) away from us—that is, at the far end of its orbit, farther away than many stars and indeed farther away than the majority of the stars visible to the naked eye.

All the arguments advanced against the passing star hypothesis showing that close passage of other stars is an insufficient cause for the perturbation of comets into earth-crossing trajectories apply equally to the Nemesis hypothesis. A further argument against the Nemesis hypothesis is that nearby stars actually pass close enough to perturb the Oort Cloud more often than Nemesis ever could, but such passing stars do not perturb comets into earth-collision orbits.

The orbit of Nemesis is easy enough to compute, and the computations also suggest that this orbit is not stable. Many other stars are nearer to the sun at present than Nemesis. Indeed, stars visible to the naked eye are rarely more than 10 parsecs away, so most of the stars we can actually see at night are at comparable distances to that of the supposed Nemesis. To return to the neighborhood of the sun within another few million years, Nemesis would have to thread its way between other stars, each one perturbing it to some extent. There is no possibility that it could have remained in any orbit near the sun for more than 250 million years.

Furthermore, as I have shown earlier, the interval between mass extinctions is stochastic, not periodic—that is, mass extinctions recur at irregular chance intervals, not at regular intervals. In other words, the very foundation on which the Nemesis hypothesis is based is fundamentally flawed.

Finally, we know of no binary pair of stars with anywhere near such a separation of the two members. In other words, Nemesis cannot exist, cannot account for the phenomena we observe, and is an unnecessary hypothesis based on erroneous statistics.

There is nothing like a good controversy and the complete demolition of someone else's arguments. All I have to do now is wait for a rebuttal in equally strong language. I love it.

Planet X

A third hypothesis implicates the supposed planet X, which has been suggested to account for irregularities in the orbit of Neptune.

After the success of Pluto in accounting for some perturbations in Neptune's orbit and its discovery near the predicted location in the heavens, some anomalies still remained in the computed orbit of Neptune. Its orbit was still thought not to be fully explained, even taking Pluto into consideration. Accordingly, a tenth planet, planet X, was suggested to account for the remaining residual anomalies.

Any conjunction of planet X with the gas giants, as suggested by this hypothesis, simply will not produce the effects observed at the Cretaceous-Tertiary Boundary. Moreover, there is no longer any reason to believe in the existence of planet X, since the supposed perturbation of the orbit of Neptune is now known to be an artifact of poor observation. Besides, the existence of planet X is incompatible with the existence of a belt of cometary objects just beyond Neptune, whose existence has been postulated to account for the short-period comets ("short" meaning less than 200 years in this context). The first such object in this zone has now been observed, and its position is incompatible with the existence of any planet X.

Giant Molecular Clouds

A final hypothesis, which has a more solid basis in celestial mechanics, is that of the passage of the solar system through giant molecular clouds. Such clouds are more abundant near the central plane of the galaxy. The entire solar system is revolving around the galactic center in the same way that the earth is revolving around the sun and the moon around the earth. Such a galactic revolution takes about 200 million years, and the sun does not just go around the galaxy but moves slowly up and down like a carousel horse, taking about 60 million years from the moment it reaches the top of its ride to the next topping out. In addition, its ride takes it sometimes through spiral arms of the galaxy, sometimes through the relatively empty spaces between these arms.

This all makes for a rather bumpy carousel ride. The solar system may be perturbed when it passes through a spiral arm or when it passes through a giant molecular cloud, the sort of cloud from which stars are formed in the first place. With a 60-million-year cycle for the up-and-down motion, the solar system would pass through the central plane of the galaxy once every 30 million years—once on the way up, and once on the way down. If we suppose that this central plane holds a giant molecular cloud getting in the way of the next descent of our carousel horse (we are near the top of this ride at the moment), then it seems reasonable that the entire solar system passes through this cloud (revolving around the galaxy at the same rate as the solar system) every 30 million years or thereabouts. The idea is that the drag of such a giant molecular cloud through which we may pass at regular intervals could perturb the Oort Cloud.

Like the foregoing hypotheses, this one suffers from problems. First, of course, is the nonperiodic (stochastic) nature of the mass extinctions; this hypothesis has been advanced to explain periodic extinctions. Second, as with the other suggested mechanisms, the chief effect of the drag of a giant molecular cloud on the Oort Cloud would be to detach comets from the solar system and send them on hyperbolic orbits into interstellar space, not to perturb them into earth-crossing orbits. And third, despite searches, we have no evidence at all that a giant molecular cloud is actually situated in the path of the descending solar system.

A Supernova as Accelerator

None of these suggested mechanisms could propel a comet into an earth impact at the speed needed to supply the energy released at the Cretaceous-Tertiary Boundary. Such a comet could only have been mov-

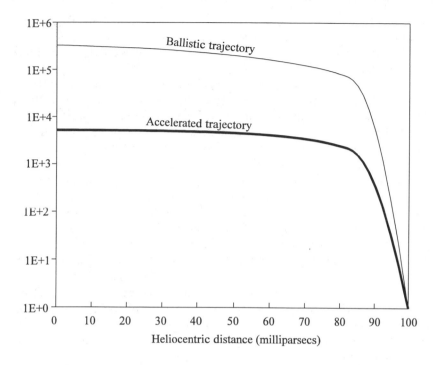

Fig. 11.2. Time course of comets in free-fall and accelerated by a supernova.

ing at a free-fall velocity, the velocity of objects moving under the influence of gravity alone. The impacting comet (or comets) of the Cretaceous-Tertiary Boundary event was moving much too fast to have been put into free-fall by any one of these mechanisms. Instead of a geocentric speed of less than 11.2 kilometers per second (or somewhat more if it had passed close to Jupiter) that these other suggested mechanisms would produce, the comet hit the earth with an impact velocity of over 80 kilometers per second. From somewhere it had picked up extra energy or momentum far in excess of gravitational energy. The only possible source of nongravitational energy or momentum is a nearby supernova. I suggest that the "great red dragon" whose fiery breath roused the comets of the Oort Cloud was indeed a supernova.

Figure 11.2 charts the time course of a free-falling versus a supernova-accelerated comet. The inner edge of the Oort Cloud lies about 100 milliparsecs from the sun. Any comets perturbed by gravitational effects into earth-crossing trajectories will move in ballistic orbits, in free-fall, slowly accelerating as they come into the inner solar system. Such a comet will take about 330,000 years to reach earth's orbit on a ballistic trajectory if the perturbing force was purely gravitational, as it would be if the perturbing force was a passing star or Nemesis. If, instead, it was accelerated by a nearby supernova, the time to arrival is cut down to about 5,000 years.

How near and how big must this hypothetical supernova have been? Detailed mathematical modeling suggests that it was a star about 20 to 25 times the mass of the sun and that it blew up about 1.3 to 1.5 parsecs away, a little farther away than the nearest star at the present day, Alpha Centauri, which lies at about 1.2 parsecs (about four light-years) from us.

In Chapter 5 I have outlined the structure of the Oort Cloud and showed that in fact it consists of two belts of comets orbiting the sun (see Figure 5.1A). The outer Oort Cloud surrounds the whole solar system, lying approximately in a distorted sphere, all around the sun, not just near the plane of the ecliptic. It is deformed into an egg-shaped cloud by the gravitational pull of the mass of the galactic center, that dense agglomeration of stars and perhaps a black hole around which the rest of the galaxy revolves.

The inner Oort Cloud, by contrast, separated from the outer cloud by a second forbidden zone and lying chiefly near the plane of the galaxy, is pulled into an oval and somewhat away from that plane by the gravitational attraction of the galactic center. Unlike the outer cloud, it does not extend too far away from that plane.

It is the comets in the inner Oort Cloud that may be perturbed by a nearby supernova into earth-crossing orbits. Figure 11.3 graphically represents the trajectories of such comets; graph A represents distances on a linear scale, while graph B represents them logarithmically. A comet perturbed into an earth-crossing orbit from the inner edge of the Oort Cloud (about 100 milliparsecs from the sun) accelerates from its initial velocity of about 100 meters per second (relative to the sun) to a maximum of about 90 kilometers per second at perihelion as it rounds the sun just inside the orbit of Mercury. The pattern of this acceleration differs between ballistic comets perturbed by gravitation effects and those accelerated by a nearby supernova, even though initial and final velocities are the same. This is particularly apparent in graph B, in which both axes of the graph are logarithmic.

Specifically, comets so perturbed are those whose orbits are sufficiently eccentric to take them between the inner and outer edges of this Inner Oort Cloud and that are at the moment of the supernova event approaching perihelion—that is, those comets just approaching their nearest point to the sun, having come in from their farthest point, the aphelion. Such an orbit takes about a million years for one complete revolution of the solar system. There are approximately two billion comets in such orbits

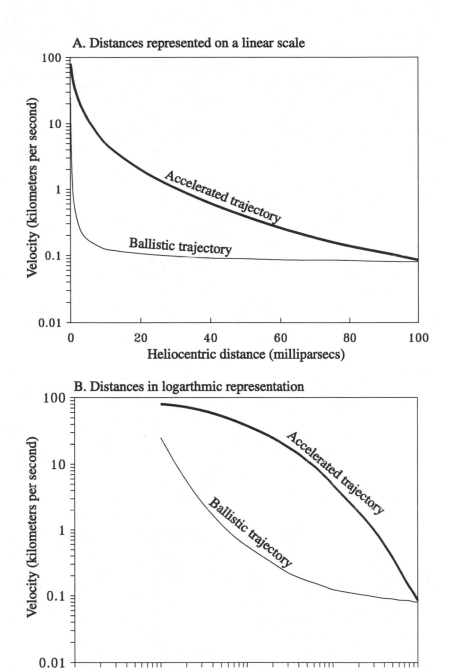

Fig. 11.3. Trajectories of comets perturbed from the Oort Cloud.

(2×10^9), and at any one moment some two million comets are at positions in their orbits where a supernova could kick them into trajectories that would bring them into the inner solar system.

What can we surmise about a supernova that might have this effect? We can reach some conclusions about the mass of the supernova and about its distance by mathematical modeling, and we have some statistics of the frequency of occurrence of supernovae, both in our own galaxy and in others. Based on this information, we may conclude that a star must have been between 20 and 25 solar masses and have been within about 60 parsecs of the sun to have had a significant effect on this sensitive group of 2×10^6 comets in the inner Oort Cloud. For example, a supernova at 50 parsecs (and we have evidence from the Vostock ice core from Antarctica that such an event occurred about 35,000 years ago at that distance) would perturb a number of comets, not into earth-crossing orbits, but into the Kuiper Belt, that unstable belt of comets located just outside Neptune.

Such an occurrence, a supernova between about 12 and 60 parsecs, seems statistically to occur somewhere between once every million years and once every three million years (stochastically, of course, not periodically). We cannot be more precise, because the statistics are poor. It seems likely to me that this is the main mechanism for the constant recruitment of new comets to the Kuiper Belt, the source of short-period comets, where comets may be placed in parking orbits for up to ten million years but not longer.

A supernova closer than about 10 to 12 parsecs (depending on its mass) can throw the susceptible comets from the Inner Oort Cloud into earth-crossing orbits. The sizes of these comets are somewhat smaller than the general population in the Oort Cloud, as Figure 11.4 shows. Such events may occur once every 30 to 90 million years, according to our best mathematical models. One as close as 1 parsec may occur every billion years, give or take a few hundred million.

Multiple Impacts

Taking into consideration the energy involved in the Cretaceous-Tertiary Boundary event and the implied heliocentric velocity of the comets, we are led to the conclusion that the best fit we can obtain between theory and observable fact is obtained when we postulate a supernova of 20 to 25 solar mass as at about 1.3 to 1.5 parsecs from the sun. This event would precipitate about two million comets into fast-moving earth-crossing orbits. These comets would be moving in close formation, and as many as a hundred of them may actually have hit the earth, totaling about

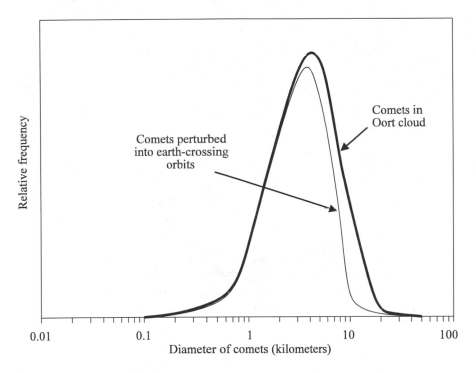

Fig. 11.4. Size of comets. Comets perturbed into earth-crossing orbits by a nearby supernova tend to be rather smaller than those in the general population of the Oort Cloud (based on the results of theoretical calculations).

2×10^{18} grams altogether, with perhaps as much as a half of this mass consisting of a single object. Multiple comet impacts seem more likely than just a single one.

There are further hints that the Cretaceous-Tertiary Boundary event involved more than one impact. Some twelve other craters, besides the one in Yucatán, are known whose date is indistinguishable from the age of the Cretaceous-Tertiary Boundary (though Alan Hildebrand of the Geological Survey of Canada now claims that he can tell their dates apart). The best-known is the Manson structure in Iowa state, a totally buried crater about 35 kilometers across. If we conceptually move the tectonic plates back to where they were 65 million years ago and draw a great circle through the Yucatán crater and the Manson structure, we find two more 65-million-year-old craters on the same circle, one in China and one in Siberia. This seems to imply that a single fragile object, a comet, broke up into several pieces on entering the earth's atmosphere on a rather flat trajectory and

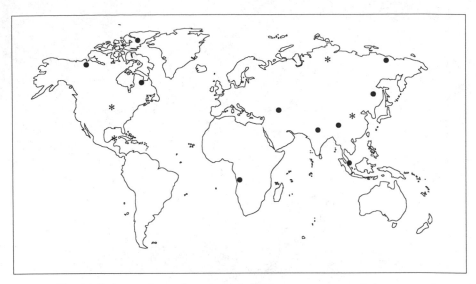

Fig. 11.5. Approximate location of the known craters of Cretaceous-Tertiary Boundary age. The asterisks show the four craters that, at the time of their formation, lay along a great circle.

that these four craters are where the four biggest of these pieces hit. Figure 11.5 shows the approximate location of known craters whose age is not distinguishable from the age of the Cretaceous-Tertiary Boundary.

And what of the other 65-million-year-old craters shown in Figure 11.5? Two possibilities exist. First, they may all have resulted from the impact of parts of a single fragile body. An approaching comet could easily break up in space even before entering the earth's atmosphere. Such events have been observed on several occasions, most notably comet Biela, which in the middle of the last century broke up into four pieces that then traveled on subparallel courses. The Yucatecan bolide and the three others that seem to have originated from atmospheric breakup could have been just parts of a single object that was fragile enough to split in this way when subjected to the tidal effects of approaching planetary masses. The other nine craters not in line with the Yucatán-Manson axis (as it was 65 million years ago) might represent other pieces of this single parent comet.

The second possibility, as noted above, is that there were multiple incoming comets. We must remember, though, that if a hundred comets were to hit the earth, as predicted by my model of solar system mechanics, then some 30 would have hit the land and 70 would have impacted the sea, the latter producing craters that would no longer be visible today. The thirteen known craters of Cretaceous-Tertiary Boundary age seem to represent a fair rate of survival among the 30 that may have been formed on land. It is thus possible that most of the craters represent independent arrivals of different comets—a shower of comets, as postulated by Piet Hut.

We simply cannot tell at the moment whether we are dealing with a hundred bits of a single comet or with a shower of a hundred independent comets arriving within a very brief time of one another. Perhaps it was a mixture of both. But whatever the mechanism, it seems certain that the Cretaceous-Tertiary Boundary event involved a number of cometary impacts, either as fragments or as whole bolides.

The Search for
Supernova Debris

One Moment in Annihilation's Waste,
One Moment of the Well of Life to taste—
  The Stars are setting, and the Caravan
Draws to the Dawn of Nothing—Oh make haste!
  —*The Rubaiyat of Omar Khayyam*,
  Quatrain XLIX
  (Rendered into English verse by
  Edward FitzGerald)

The whole solar system was formed by condensation from a preexisting cloud of gas compressed by the shock waves from supernova explosions. I have made this statement earlier in this book, but it bears repeating.

The molecular cloud from which the solar system, and indeed the whole star cluster, condensed was not primordial in any way but contained a great deal of debris from an earlier generation of defunct stars. Indeed, the cloud of gas certainly contained the remains of second-generation stars; our sun is in fact a third-generation star.

Some Background on Isotopes and Stars

All elements and isotopes heavier than iron, of atomic mass 56, were formed in supernova events. Iron 56 is the most tightly bound and hence least energetic of all atomic nuclei, and its formation from lighter elements releases energy; but so does its formation by fission of heavier elements. We can understand this best if we consider the release of energy by the fission of heavy elements. This is the principle of the atom bomb, which uses uranium of mass 235 or plutonium of mass 238 as the nuclear fuel. Fission of these isotopes to form lighter isotopes (for instance, iron 56) releases the killing energy of the atom bomb.

The hydrogen bomb works on the opposite principle, releasing energy by the fusion of two light atomic nuclei to make heavier isotopes, with masses up to about 250, though most of them are below 56. There is an energy minimum at iron 56, so that fission of isotopes of greater mass than this will release energy—as in the atom, or fission, bomb. Conversely, fusion of isotopes of lower mass than 56 also releases energy. The fuel of a star is light elements, and a star shines by nuclear fusion, the nucleosynthesis of heavier isotopes (up to iron 56).

Inside a star, then, which radiates as a result of fusion of its light nuclear fuel, only isotopes up to iron 56 can be formed. Hydrogen-to-helium fusion releases energy; helium to lithium to carbon to iron are all fusion processes that release energy, which is radiated off into space as the light of the shining star. Any fusion beyond iron actually uses up energy. This radiation of energy produced by atomic fusion prevents a star from collapsing in on itself by its own gravity and enables it to continue shining. The size of a star is a balance between the centrifugal force of radiation pressure and the inward pull of self-gravitation. So long as there is sufficient nuclear fuel to maintain the radiation pressure, the star remains essentially constant in size. I say "essentially constant" because on a small scale the sun, and any star, is pulsing in and out like a beating heart. But once the fuel is used up to the point that radiation pressure is no longer enough to counter self-gravity, then the death throes of the star begin.

The form that the death throes take depends in part on the mass of the star and also on whether it is solitary or a member of a binary system of two stars. A solitary star with the mass of the sun, for instance (and remember that the sun is solitary, not partnered by the hypothetical Nemesis), would begin to expand as soon as the final fuel in the very core is exhausted and only the outer layers continue to produce radiation. The sun itself, in about five billion years' time, will expand to become a bloated red giant, expanding beyond the orbit of the earth. The world will indeed end in a bang as a puff of vapor. With the final exhaustion of most of its fuel, radiation pressure will no longer be able to sustain the red giant against its own gravity, and the star will shrink to become a white dwarf.

A star three or four times the mass of the sun will shine more brightly and use up its fuel more quickly, so even though it starts with more nuclear fuel it has a shorter life. A star of this mass cannot retain its integrity through the final death throes and, after the red giant phase, will collapse more abruptly and then blow itself apart into a nova, dispersing all of its mass. Part at least of the cloud from which the sun was originally formed four and a half billion years ago was supplied from the remnants of one or more novae, which supplied most of the iron found in the inner stony planets, including the earth.

An even more massive star, say 20 to 25 times as massive as the sun, lives for an even shorter time, burning up its nuclear fuel yet more quickly. It too will become a red giant toward the end of its life, but the final collapse is much more dramatic. From a red giant almost the size of the whole solar system (except for the Oort Cloud), bigger indeed than the planetary system, it collapses in a few seconds to a knot smaller than the moon. Under the hammer blow of 25 solar masses hitting this center at a velocity of hundreds of kilometers a second, the core rings like a bell, and about 1.2 solar masses of central material is converted into neutrons and neutrinos, one neutrino for each neutron. A rebound follows, the neutrinos are radiated off into space, and the resulting explosion pushes off the rest of the star's mass at a speed of thousands of kilometers a second. The remaining central core quickly becomes a neutron star, with more mass than the sun in a volume no more than 20 kilometers across, or perhaps it becomes a black hole. About two solar masses of the expelled matter is converted to energy (in addition to the energy carried off by the neutrinos), some of which is radiated into space. Much of the energy is absorbed by the expelled matter in energy-absorbing nuclear fusion reactions to form atomic nuclei of high mass, above the energy limit of iron 56. So far as we know all isotopes, stable as well as radioactive, with mass over 56 are formed in this way when a star goes supernova.

Why am I writing about all this? It is the work of other scientists; I have contributed little to the ideas expressed here. I have oversimplified with reckless speed and have totally neglected the death throes of binaries. My only excuse is that it might help you to understand why and how I embarked on the thrill of the search for supernova debris.

Searching for Supernova Debris

The primordial cloud from which the sun and the solar system developed certainly contained nova debris and also supernova debris—or else the earth would have no heavy elements—no lead, no iodine, no gold, platinum, or silver. In this context, heavy elements are those with an atomic mass above 56, though astrophysicists sometimes call all isotopes above helium, of mass 4, "heavy elements" or even "metals." To them carbon is a heavy element and a metal! More supernova debris must have been added to the cloud in the very formation of our solar system, which we believe took place when the shock wave from a supernova (or several supernovae) initiated collapse of the cloud into self-gravitating nuclei.

We are now faced with a dilemma. If, as I am suggesting, a supernova was somehow involved in the Cretaceous-Tertiary event, how are we to search for supernova debris from that particular event in the face of much more abundant debris left over from the formation of the world and of the whole solar system? In fact, we have both conceptual problems and also technical problems.

There are two possible lines we can follow. One is to search for a stable isotope that is abundantly formed in a supernova, either directly or indirectly, and hope that this will introduce an anomaly in isotope ratios. Ordinary silver, for example, is a mixture of two stable isotopes, silver 107 and silver 109, in roughly equal proportions. Silver 107 is the daughter product of the radioactive disintegration of palladium 107, and an abundant product of a supernova. An anomalously high level of silver 107—a mere one part in a hundred thousand or even one in a million—could be a marker of supernova activity. We can think of no other mechanism by which such an anomaly might be developed.

The second line of investigation is to search for certain radioactive isotopes formed in a supernova and nowhere else. This limits us to isotopes of mass greater than 56. There are also other severe restraints on what we might usefully seek. The first restraint is that the half-life of the isotope should be long enough for detectable amounts to have survived the 65 million years since the Cretaceous-Tertiary Boundary event. In effect this limits us to isotopes with a half-life over about fifteen million years, so that no more than four half-lives have elapsed since the Cretaceous-Tertiary event.

The second restraint is that the half-life should not be long enough for the isotope to have survived from primordial times, 4.5 billion years ago. Thus uranium 238, which has a half-life of about ten billion years, could not serve as a marker for a Cretaceous-Tertiary supernova even though it was certainly formed in supernovae; most of it probably came from the original condensation of the gas cloud from which our solar system was formed. In practice, we are restricted to radioisotopes with a half-life no more than a tenth of the lifetime of the earth, say 450 million years.

The third constraint, an obvious one, is that the isotope should not be a fission product, directly or indirectly, of one of the primordial radioactive materials such as uranium or thorium. This constraint would rule out radium, for instance, or radon, both daughters of uranium produced by fission of that element. In fact, only proton-rich isotopes are possible candidates for markers for supernova involvement in the Cretaceous-Tertiary Boundary event—that "Moment in Annihilation's Waste."

Within these restraints, then, we find that only four radioisotopes are possible candidates as supernova debris at the Cretaceous-Tertiary Boundary: iodine 129, with a half-life of 17 million years; uranium 236, with a half-life of 16 million years; plutonium 244, with a half-life of 82 million years, and curium 247, with a half-life of 16 million years.

A further practical constraint is the possibility of contamination of surface rocks with these isotopes by fallout from hydrogen bomb tests or from nuclear accidents such as Chernobyl. Strict cleanliness in collecting samples, comparison between surface-exposed samples and samples from deep within the rock face, and comparisons between different horizons are all needed in this search for possible supernova debris. In addition to anthropogenic, or man-made fallout, iodine 129 is formed naturally by spallation (literally, whittling away) of barium by cosmic rays in the atmosphere. There is thus a background level of this isotope in all the rocks. We need to seek for an anomalously high level of iodine 129, not its mere presence.

In the face of all these constraints, the candidates for possible supernova markers are the following

Silver 107	anomalously high level, evinced in silver isotope ratios
Iodine 129	anomalously high level, significantly above background
Uranium 236	presence in the boundary layer, absence elsewhere
Plutonium 244	presence in the boundary layer, absence elsewhere
Curium 247	presence in the boundary layer, absence elsewhere

Technically, measuring any one of these five presents difficulties and involves instrumentation that is pushing the envelope of current sensitivity.

In a sense, the silver isotope measurements are the easiest, demanding nothing more elaborate than a sensitive mass spectrometer coupled to some kind of separation device, a chromatograph or an inductively coupled argon plasma source (ICAP). This may sound simple, but we are now pushing the limits of sensitivity. The smaller the amount of material we need to measure, the greater the precision we desire; the lower the concentration we wish to analyze, the larger is the equipment we need and the higher the energy level required, until we are using not just rooms full of machinery but equipment that won't even fit into a normal-sized football stadium. Fortunately, silver isotope ratios can be measured using equipment small enough to fit into a large laboratory.

The radioisotope measurements demand even higher-energy equipment than do the silver isotope measurements and can only be carried out by the use of accelerator mass spectrometry. Gamma-ray counting will not suffice; there are just not enough atoms of interest to expect even one to disintegrate during the course of a count. We might just as well seek a hair from the beard of the Great Cham. A mere handful of accelerator mass spectrometers are in operation around the world, and only the University of Toronto machine, located in the IsoTrace Laboratory (Figure 12.1), is at the moment actually capable of making the measurements required for this search.

Accelerator Mass Spectrometry

The Toronto accelerator mass spectrometry machine operates at three million volts and consists of a low-energy (120 kilovolts) mass spectrometer that cleans up the initial beam of negative ions, rejecting ions whose mass is too far from the isotope under investigation. Then comes the tandem accelerator, with a central electrode maintained at a positive potential of up to 3 megavolts. At this point, near the positive electrode, excess electrons are stripped off the particles in the beam, converting them to positive atomic ions and splitting any molecular fragments that may have been in the initial negative ion beam. The ions are now repelled by the positive electrode and accelerated to the far end of the tandem. The beam emerging from the tandem particle accelerator is then passed to a second mass spectrometer for final analysis, which involves the counting of individual atoms of the desired isotopes. This is done with magnets weighing as much as 15 metric tons for heavy elements such as those we find of interest in this research. Figure 12.2 provides a schematic diagram of this process; the radiocarbon line shown in this schematic is not used in the analysis of heavy elements.

Ted Litherland, Linas Kilius, and Zhao Xiao-Lei have collaborated with me in this search for supernova debris at the Cretaceous-Tertiary Boundary. Ted Litherland is director of the IsoTrace Laboratory, Linas Kilius has spent his whole research career there, and Zhao Xiao-Lei had just completed his Ph.D. thesis (on negative ions of radium) there when he began this collaboration. As for me, I was for several years chairman of the board of directors of IsoTrace.

Accelerator mass spectrometry was initially developed for carbon-14 dating, for which it is the most sensitive, most precise and most accurate method available. At the time of writing there are only seven fully operational instruments in the world used regularly for accelerator mass spectrometry, and six of these are used only for carbon-14 studies, since they lack a heavy element line. The exception is the Toronto facility, which is

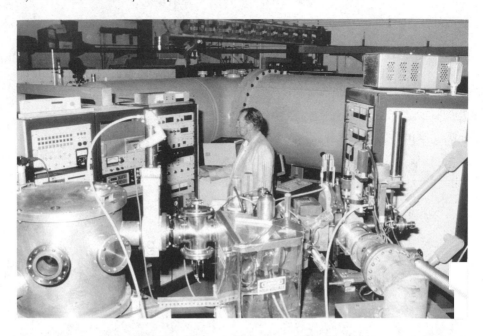

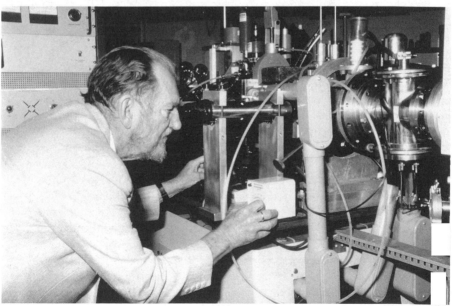

Fig. 12.1. The accelerator mass spectrometer in the IsoTrace Laboratory at the University of Toronto, with the author.

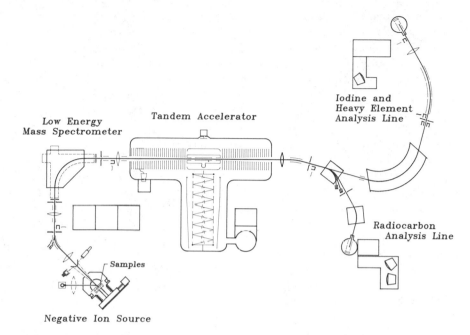

Iodine and
Heavy Element
Analysis Line

Low Energy
Mass Spectrometer

Tandem Accelerator

Radiocarbon
Analysis Line

Samples

Negative Ion Source

Fig. 12.2. Schematic of the accelerator mass spectrometer in the IsoTrace Laboratory of the University of Toronto, courtesy of Dr. W. E. Keiser.

used for a varied program of research of many kinds on other elements, as well as for carbon-14 studies. In addition, there are a number of other accelerators sometimes used with spectrometers, and several more in various stages of development, almost entirely for carbon studies.

The heavier elements require somewhat different peripheral equipment if they are to be studied by accelerator mass spectrometry. Much of this equipment has been designed in Toronto. Let me first describe how the machine is set up, both in Toronto and at the other laboratories, for carbon-14 studies.

The carbon samples, generally about 1 to 10 milligrams, are placed in small cylindrical depressions in a sample holder/changer. Each sample holder has a number of sample pits that are used to hold duplicate samples, together with reference and standard materials of known composition, which are used for calibrating the instrument. Once the sample holder is placed in the machine, the air from the chamber holding it is pumped out until the pressure reaches the high vacuum in the rest of the machine. This chamber is then opened to the beam line. From now on everything is computer controlled, including changing from one position on the sample holder to another.

The samples (and standard materials for comparison, in succession) are maintained at a steady negative voltage of 120,000 volts and bombarded with a finely focused beam of cesium ions, a process called sputtering. This knocks negative carbon ions off the surface of the sample—that is, it knocks off carbon atoms and gives each one an extra electron, making it negatively charged (C⁻). Unfortunately, sputtering also knocks molecular fragments off the samples, such fragments as negatively charged carbon hydride (CH⁻).

Since all these particles are negatively charged and the samples are at a negative potential of 120,000 volts, they are repelled by the sample. (Like charges repel and unlike charges attract one another.) The resulting stream of negative ions is focused into a beam that is injected into a low-energy mass spectrometer. Here the beam is bent by a combination of large electromagnets and electrostatic separators. The degree of bending depends both on the charge and on the mass of each ion. Any ions that are outside the range of interest (for carbon studies this is a single-charge state and a mass between 12 and 14) are rejected from the beam; they are either bent too much or too little to pass through the exit slit and are absorbed in so-called Faraday cups.

The beam that emerges from this stage of the machine is too powerful to be injected in its entirety into the next stage of the machine, the tandem accelerator, so the power to the main low-energy magnet ("low" because the operations at this stage are at a mere 120,000 volts) is switched cyclically so that only ions of one particular mass are entering at any one moment. The degree of bending of the beam is changed slightly, hundreds of times a second, so that in succession ions of mass 12, mass 13, and mass 14 are injected into the accelerator more or less separately. The low-energy spectrometer is not capable of total separation of these masses, so relying only on this part of the instrument (as is done in conventional mass spectrometry, which does not use an accelerator) is inherently inaccurate.

The accelerator is in two sections, with both ends maintained at ground potential (zero volts). At the center is an electrode that is maintained at three million volts positive potential. The negative ions are attracted toward this electrode and accelerated through the vacuum of the accelerator chamber. Just before they reach this positive electrode they pass through an electron stripper. In the Toronto machine this consists of a wispy cloud of argon gas contained in a sort of "bottle" to prevent it from dispersing into the vacuum of the main chamber. As the negative ions and molecular fragments collide with the argon atoms, excess electrons (as many as four of them) are stripped off and molecular fragments broken up. The ionic particles now carry a positive charge, a triple positive charge, in fact, since they have lost not just the single extra electron that gave them a negative charge, but three more as well. Their momentum has carried them past the central positive electrode of the accelerator, which now repels them toward the far end of the chamber.

The total gradient over which the ions have been accelerated is thus about six million volts, some three million in their initial attraction as negative ions to the three-million-volt electrode in the center of the chamber and another three million volts after they have been stripped of electrons to form positive ions. We must subtract about 300,000 volts from this acceleration for the loss of momentum when they hit the electron stripper and add 120,000 volts for the acceleration in the initial low-energy end.

The emerging beam, now cleaned of all molecular fragments, is injected into the high-energy mass spectrometer. Here final separation of the three isotopes of carbon—carbon 12, carbon 13, and carbon 14—is achieved. Each enters its own separate detector. The rare carbon 13 and the even rarer carbon 14 are counted, atom by atom (or rather ion by ion), while the abundant carbon 12 is measured as a current. A carbon 14 date is obtained from the ratios of these isotopes.

To adapt the machine for heavy element analysis, several changes are requisite. First, a much bigger magnet is needed for the high-energy mass spectrometer. In the IsoTrace Laboratory in Toronto, the magnet we use weighs 15 metric tons. Second, sputtering, while it is certainly the most efficient way of creating negative carbon ions, is not always the best way of creating a stream of heavy ions. For the moment, though, we are confined to the use of cesium sputtering even for heavy elements. The low-energy end of the machine too needs modifying, at least for the higher energies needed to bend a beam of heavy ions, and, as Zhao Xiao-Lci has shown, other changes too are needed in order to achieve the highest precision.

The Isotrace Laboratory had its first successes in heavy element analysis with isotopes of iodine, and one of these, iodine 129, is the first isotope we have investigated in the Cretaceous-Tertiary Boundary rocks in our search for supernova debris.

Early Results on Supernova Debris

The first positive find in our search for supernova debris was iodine 129, the work being carried out largely by my colleague Linas Kilius. The coal immediately above the Cretaceous-Tertiary Boundary, or rather the lowest 10 centimeters of this coal, had levels of this isotope about twenty times the background level found in other adjacent rocks. It was, however, in the coal layer, not in the Cretaceous-Tertiary Boundary clay itself. This is perhaps not unexpected, for the tree ferns, whose remains form the coal, had roots penetrating into the Cretaceous-Tertiary Boundary layer, and iodine is biologically mobile. In fact, what was measured was not the absolute level of iodine 129 but the ratio of its abundance to nonradioactive iodine 127. The overall level of iodine varies enormously in rocks because of the activity of biota in absorbing iodine in general. Coal seams are all rich in total iodine, having been enriched in this way during life. What was unique about the lowest 10 centimeters of the coal lying just atop the Cretaceous-Tertiary Boundary layer was that it was enriched in iodine 129, relative to the already high enrichment in iodine generally. It therefore seems likely that the Cretaceous-Tertiary Boundary claystone was initially enriched in iodine 129 and that this isotope has now migrated into the coal through absorption by the roots of tree ferns. The ratio of iodine 129 to stable iodine is about twenty times that found in nearby Cretaceous or Tertiary rocks. In other words, the coal immediately overlying the Cretaceous-Tertiary Boundary is enriched twentyfold in iodine 129.

The second result was finding an anomalously high silver-107-to-silver-109 ratio (high by about one part in 4,000) in the Cretaceous-Tertiary Boundary clay itself, compared with rocks a few centimeters above and below. This element is not taken up by tree roots, and there is no reason to suspect that it would migrate into other layers from its site of deposition. This additional silver 107 can only be the daughter of the supernova product palladium 107.

In the search for plutonium 244 (which is not the atom bomb isotope of plutonium) and for curium 247, we are searching for little more than single atoms. Zhao has joined me in developing the techniques required for this search and has succeeded in showing for the first time that both these elements possess stable negative ions, something that is necessary for accelerator mass spectrometry. He has also shown for the first time the presence of uranium 236, which until now has been a laboratory curiosity. For my part, I have developed the methods needed to concentrate the slight traces of these isotopes from large quantities of rock into samples small enough to be used in the accelerator mass spectrometer. Since the elements in question exist in the rocks in the form of refractory oxides, which cannot be dissolved in acids, the first step is to dissolve all the other minerals. A sample of rock weighing about 200 grams yields about 10 milligrams of refractory oxides, which we are using for identification first of uranium 236, then of plutonium 244, and curium 247. Our first results with plutonium 244 and uranium 236 are suggestive but fall short of the levels of proof required by scientists.

For the moment, we can claim to have found unequivocal supernova debris in the rocks associated with the Cretaceous-Tertiary Boundary in the form of the silver isotope anomaly and in the iodine 129 anomaly. This forms a first test of my model of celestial dynamics, which predicts that the comet (or comets) that hit the earth at the time of the Cretaceous-Tertiary Boundary event were accelerated by a nearby supernova.

Why do we find soot just below the Cretaceous-Tertiary Boundary layer, silver isotope anomalies, extraterrestrial diamonds, shocked quartz grains, even plutonium 244 and uranium 236 in the Cretaceous-Tertiary Boundary claystone itself and an iridium anomaly and excess iodine 129 just above the boundary? And what does all this have to do with the demise of the dinosaurs? We shall find out as I endeavor to tie together all the strings and provide a denouement to this celestial detective story.

13

A Scenario of the Cretaceous-Tertiary Boundary Event

When beggars die, there are no comets seen;
The heavens themselves blaze forth the death of princes.
—William Shakespeare, *Julius Caesar*

There were no beggars at the end of the Cretaceous. The heavens blazed forth, and almost all living things on the surface of the earth died. Comets appeared in the skies. Did I say comets? Plural?

Have you ever thought how many stars you can see in the sky with the naked eye? How many you see depends on where you are and the state of the skies. Clouds may obscure the sky. The moon may shine so brightly that only the brightest stars may be visible. A recent volcanic eruption, such as Pinatubo or Mount Rush in Alaska, may cast a pall of dust across the sky so that even on a clear moonless night many fainter stars become invisible. Streetlights too make an enormous difference in how many stars we can see even on a clear moonless night. In the center of a big city—perhaps San Francisco, New York, or London—it would be rare to be able to see even as many as two hundred stars on a clear night. In the London of Queen Elizabeth I, when there were no streetlights and people kept their windows tightly shuttered at night, the poet and playwright Kit Marlowe had it right in *The Tragical History of Doctor Faustus* (1604):

Oh, thou art fairer than the evening air,
Clad in the beauty of a thousand stars.

A thousand stars is about right for the London of his day or for the more poorly lit suburbs today, where streetlights are few and far between. On a clear night in the country in summer at a lakeside cottage, lazing on the dock at night, perhaps with a drink in your hand, you may count twice that number. The humidity in the air absorbs a surprising amount of starlight. Go to that dock in winter, when the lake is frozen over (I am writing from Canada, not Lake Okeechobee in southern Florida) and the air is drier with the cold; then you may see as many as three thousand stars on a cold, dry moonless night. Drier still, in the Butana desert of Africa, I have estimated as many as four thousand stars visible on a night when the needle on the humidity gauge would not leave the pin. Four thousand stars is just about the maximum number visible to the naked eye under the best of seeing conditions, stars exceeding magnitude 6 in brightness.

A Review

Let us now weave all the strands, both observational and theoretical, together to present a scenario that is consistent with all we have observed and with all we can calculate. I have argued in Chapter 10 that a nearby supernova was the prime cause of the Cretaceous-Tertiary Boundary event. This supernova perturbed the Oort Cloud by transferring momentum to those comets that were approaching perihelion and whose trajectory at that moment was directly away from the supernova. Roughly two million comets were at this point in their orbits when the supernova explosion occurred. All the comets in the Oort Cloud were perturbed and their orbits changed, but only those at this particular point in their orbits were actually shoved out of the cloud and accelerated toward the sun into earth-crossing trajectories.

Imagine now two million comets streaking in toward the sun at roughly the same speed as a result of this bang. It will take them about five thousand years to reach the inner solar system, by which time they will have spread out, some moving a little faster than others. Some will have passed close to a gas giant, such as Jupiter, which would change their orbits yet again, but most of them will cut into the zone of the stony planets, make a quick left turn (in a prograde orbit they are moving counterclockwise as seen from the north pole of the solar system) just outside the orbit of Mercury, and disappear forever, leaving the solar system entirely on hyperbolic orbits.

Within a period of a few centuries at most, some two million comets cross the orbit of the earth. Fifty thousand of these should have been visible to the naked eye of a primitive mammal on a clear late Cretaceous night, more than ten times as many as the visible stars. But I get ahead of myself.

I have argued in Chapter 6 that vulcanism is not adequate to account for the events at the Cretaceous-Tertiary Boundary; only some kind of collision with an object from outer space would explain what we've found. We know that many kinds of such objects, large and small, have hit the earth during its long history.

First, we have bits of asteroids, mostly in the form of meteorites but some much larger. The biggest of these of which we have direct evidence in the geological record is perhaps the Sudbury Basin in Ontario, Canada. This represents the remains of the impact site of an asteroid fragment that collided with the earth some 1.82 billion years ago. A part of the metallic core of an asteroid dealt the earth a glancing blow, making an elongated crater, though part of this elongation may be a result of subsequent structural deformation. The chunk of metallic material, an overblown "iron" meteorite over 100 kilometers across, perhaps even over 200 kilometers, and with a mass of at least 10^{21} grams, created a great puddle of molten rock so large that it took about a million years to solidify again. During this period the heavier elements in the melt, chiefly nickel, settled to the bottom and crystallized out as a sort of sediment at the base of the melt. This is the rich nickel deposit, shaped like a giant saucer, for which Sudbury is famous. The lighter silicate rocks floated up to the surface and formed a slag there, solidifying first. Finally, the center of the mass of rock solidified, containing other mineral types and forming the overburden to the Sudbury mines. In the absence of any but rare single-celled fossils from this period, we cannot make any real estimate of the mass extinction that must have resulted from this event, but the energy involved was far greater than in the Cretaceous-Tertiary Boundary event, and it seems likely that only a few primitive photosynthetic bacteria survived.

Second, the earth has been hit by meteorites derived from the moon and Mars, thrown out from these bodies when they were themselves hit by a bolide. We have no evidence for anything larger than a small meteorite, an object not big enough to make a crater.

A third possible category is that of the Apollo bodies, which are probably dead comets in highly eccentric orbits around the sun. They are dead in the sense that any volatile materials have been boiled off in the course of many passages near to the sun. We have no real evidence that any one of these has ever in fact collided with the earth, though it does not seem improbable.

The final category is that of comets and of cometary debris and dust. Every year the earth passes in August through the orbit of a comet that has shed a cloud of particles. Until 1992 this comet, P/Swift-Tuttle, had not been seen by astronomers for over a century. Detailed calculations of its orbit since it has been sighted again have suggested to some astronomers that it has a good chance of hitting the earth some time early in the twenty-second century. Others, including myself, regard this as extremely improbable.

Some of the particles shed by the last passage of P/Swift-Tuttle, enter each year into the upper atmosphere, where they burn up as shooting stars, producing the Perseid meteor shower. Larger particles from this trail of dust penetrate through the earth's surface as meteorites, arriving as carbonaceous chondritic meteorites.

"Carbonaceous" means that they are rich in carbon and carbon compounds; "chondritic" means that they contain chondrules—minute, near-spherical objects that appear to be tiny frozen droplets of once-molten rock. Together, these materials represent the "dirt" of the "dirty snow-ball": the parent comet; the ices, both frozen water and ices formed of other compounds such as carbon dioxide (dry ice) or methane, have boiled off from the heat of the sun. The chondrules are probably particles of interstellar dust that agglomerated into the formation of the parent comet.

Carbonaceous chondritic meteorites represent about 5 percent of the total meteorites that have been recovered and are important for the information they supply about the chemistry of comets. The largest well-studied member of this group is the Murchison meteorite, with which I have (in Chapter 8) compared the chemistry of the Cretaceous-Tertiary Boundary claystone. The composition of this Cretaceous-Tertiary Boundary rock seems to be closest to a mix of about 8 to 10 percent Murchison meteorite and about 90 to 92 percent altered bedrock. The total mass of the material thrown into the upper atmosphere from the crater was about ten to twelve times the mass of the impacting bolide, so that as the dust settled the ratio of bedrock to meteoritic dust should be about twelve to one, agreeing with the figure derived from the composition of the Cretaceous-Tertiary Boundary claystone.

I must emphasize that there is no evidence at all that every mass extinction was the same or that they were periodic in any way. Almost the only thing they had in common—or at least that many of them had in common—is that they were caused by some kind of impact with an object from outer space. The mass extinction that must have occurred 1.82 billion years ago, the Sudbury event, was caused by an impact with an iron meteorite, a giant impactor, part of the core of an asteroid composed in large part of iron and nickel. The Cretaceous-Tertiary Boundary impactor was not an asteroid but a comet (or a shower of comets or cometary fragments). Bolides that caused other mass extinctions may have been comets, but could just as easily have been asteroid fragments, either metallic core fragments or stony mantle pieces. We have evidence for fifteen mass extinctions during the 570 million years of the Phanerozoic, the largest of which is that terminating the Paleozoic era. Where these events are associated with an iridium anomaly, we can be certain that the impactor was not a mantle fragment (since the mantle of an asteroid, like the mantle of the earth, is impoverished in iridium) but rather either a comet or a core fragment of an asteroid. We have no indication about the speed at which any of these impactors hit the earth and hence no idea about the cause of the initial perturbation. Mass extinction events that are not associated with an iridium anomaly may well have been caused by the impact of a stony mantle fragment of an asteroid.

Besides the energy release resulting from the impact itself, there is little these events have in common. A comet would have roughly the composition of the primordial gaseous nebula from which the solar system condensed. An asteroid lacks lighter elements, especially hydrogen, but its core is enriched in platinum-group elements, most notably iridium. The differentiation of an asteroid, however, during its early evolution, will have resulted in the segregation of these siderophilic elements into the core, leaving the mantle depleted in heavy metals. Accordingly, iridium remains at primordial concentrations in comets, is enriched in iron meteorites (derived from asteroidal cores), and is depleted in stony meteorites (derived from the outer mantle of asteroids).

As a result, an anomalously high iridium level in those rocks marking a boundary and an extinction event can only indicate that the impacting object was either a metallic iron meteorite of asteroidal origin, or else a comet. The presence of an iridium anomaly precludes any possibility that the impactor was a fragment of an asteroidal mantle or crust. Thus the Cretaceous-Tertiary event cannot have been the impact of a stony meteorite.

Some other geological horizons associated with mass extinctions show an iridium anomaly; such horizons cannot be the results of the impact with a stony meteorite. Any iridium anomaly indicates that either an iron meteorite or else a comet hit at that time; in the absence of other evidence, we cannot decide which. Accordingly, the scenario presented here refers specifically and uniquely to the Cretaceous-Tertiary Boundary event and cannot be extrapolated to cover other mass extinctions, which may or may not have happened by the same mechanism.

In particular, although we may have evidence of an impact event at another horizon, we cannot assume that this is one of a sequence or that the same mechanism threw the impactor off its original trajectory and onto a collision course with the earth. I have argued that a nearby supernova was the original perturbing agency, causing a comet (or comets) to hit the earth at the time of the Cretaceous-Tertiary Boundary, but there is no evidence whatsoever that this is true of any other mass extinction event. It may be so, but we have found no evidence for it yet. Again, the scenario presented here is specifically for the Cretaceous-Tertiary Boundary, not for any other extinction horizon.

A Present-Day Impact Scenario

In considering the effects this scenario might have had on mass extinctions at the Cretaceous-Tertiary Boundary, it would perhaps be helpful to contemplate what might happen to the flora and fauna today if a similar event occurred—for instance, if comet P/Swift-Tuttle hits the earth in another 125 years, as some astronomers have projected.

The Seas

The initial flash from a supernova, comprising visible and near-visible wavelengths (infrared, visible, and ultraviolet light) would have relatively little effect on sea life. Certainly the littoral zone (the zone of the seashore between tide marks) would be baked, the tidepools would heat up and dry out, and most of the creatures living there would be killed. But few of these plants and animals are exclusively intertidal; individuals of almost all of them live below low tide also. Even those species that are exclusively littoral as adults spend part of their life history in the sea, as larvae, spores, or what-have-you, and so would be unlikely to be wiped out as species.

Excessive light in the photic zone might drive some pelagic creatures deeper for a short time, but this is hardly fatal, and the excess levels of illumination last for so short a time anyway that this effect would be minimal.

The biggest effect of the flash would be in terms of the clouds of soot from the forest fires on land, both in the effects these might have on the light reaching the sea and in the toxic effects of runoff in estuarine and coastal zones. Altogether it seems unlikely that the initial flash would lead to the extinction of any marine species at all.

The effects of cometary bombardment would be more serious. Months of light deprivation from the stratospheric dust cloud of ejecta from the craters would result in lowered temperatures, failure of photosynthesis, and the deaths of many creatures in the sea. This might include the extinction of entire species, especially of short-lived phytoplankton. In addition, the zooplankton, which feed on the phytoplankton, and the fishes, which feed on the zooplankton (and all the way up the food chain), would suffer from lack of food. Would this be significant in terms of extinction of species? I doubt it. Certainly large numbers of individuals would die of starvation, but there seems to be no mechanism here for total extinction of many species.

More effective in killing of whole taxa would be acidification of the oceans. The shock wave from the passage of comets through the atmosphere causes the dissociation of the oxygen and nitrogen molecules of the air and the resultant oxidation of nitrogen, the production of nitrogen oxides, and ultimately, after reaction with water, the generation of large amounts of nitric acid.

Acid rain falling on land and sea would lead to the gradual acidification of the oceans, or at least of the surface mixing zone—that is, the photic zone of the ocean, the top 200 meters. Computer models suggest that this acidification would reach its peak about a century or more after a cometary impact, and the whole mixing zone would become ten times as acid as normal. In the usual jargon, a scientist would say that the pH had dropped by a full unit. This acidification, or drop in pH, is enough to prevent any creature that possesses a lime skeleton from depositing calcium carbonate to form a skeleton at all. Such creatures living in the mixing zone, which is coextensive with the pelagic zone, would be totally unable to form any kind of shell or skeleton. The chalk-forming organisms in the open ocean would die out. The pelagic molluscs would be unable to complete their life cycles. Even pelagic fish would suffer, though their skeletons consist largely of the more acid-resistant calcium phosphate, because the energy required to extract the necessary materials from the sea becomes far greater as the acidity of the ocean increases.

Inshore the clams might suffer, though not to the same extent. The abundance of limey deposits near the shore—including limestones, dead shells, coral reefs, and rocky outcrops—would largely prevent the drop in pH and the acidification of the inshore waters. Corals would survive, at least in small numbers, as would sponges, fish, and squid. No doubt the great majority of individuals would die, but enough might survive to perpetuate the species. Oyster beds too would form a refugium against the acidification of the sea, and numerous creatures could survive there. Granitic coasts, such as the Outer Hebrides, Maine, or Newfoundland, already short of calcium, would suffer most.

Planktonic species with high calcium demands—foraminifera and coccolithophorids—would be the hardest hit, though benthic species of these same groups might well survive the cataclysm. Groups with low calcium demand—diatoms, dinoflagellates, and radiolaria (Figures 3.3, 3.4, and 3.5)—would fare much better. Diatoms and radiolaria have siliceous skeletons, which are more acid resistant, and diatoms and dinoflagellates have overwintering spores that are dormant for several months on the bottom of the sea.

Coastal Lands

Coastal lands, and especially coastal islands, must be distinguished from the continental land masses, for their fate in the face of cometary impact could be quite different. I think of my own dear Outer Hebridean island where I spent so much of my youth as I contemplate what might happen during something like the Cretaceous-Tertiary event.

The initial flash would set the heather alight; there are no trees to burn, and the boggy places are too wet and mossy to ignite. This is by no means deadly to the heather, which on well-managed grouse moors is regularly burnt as a management practice, while in a state of nature regular fires are caused by lightning. The whole heather community has evolved to withstand just this kind of thing. (I am here using the term *community* in the ecologist's sense, meaning all those animals and plants that live together as a stable and cohesive ecosystem.)

This points up one factor in long-term evolution, which will be addressed in the final chapter of this book. Creatures cannot evolve special features that might be appropriate to a set of circumstances occurring only many lifetimes apart, but if they already possess characters that preadapt them to these rare circumstances, these characters may prove advantageous and permit survival. But more about that later.

The result of the supernova flash on a Hebridean island may well be negligible, since the heather community has evolved in the face of repeated fires. The burning of the heather may seem horrendous to me when I think back on the part it played in my childhood, but the community would survive. And there are no trees on the open moor.

If I were at my grandparent's home today in the Outer Hebrides, what would be the first effect I might notice from the impact of a comet at the opposite side of the ocean? Probably a tsunami, a great tidal wave punched up by the impact. Even across the Atlantic ocean (indeed, even in the Pacific), the tsunami would be overwhelming, horrendous, immeasurably grander than any tsunami in the history of mankind. Given an impact site on the Yucatán peninsula (the site of the largest of the Cretaceous-Tertiary Boundary cometary fragments), the tsunami could be expected to rise about 150 meters up the Hebridean Islands—not even considering the height of breaking waves. Once this tsunami broke over the coast, seawater would be thrown over the mountainous backbone of the islands, even the 800-meter Cuillins of Skye, which hold such a place in Scottish lore.

Surprisingly, such a tsunami would impose its major effects by scouring the land. Its drawback, or recession from the land, would suck soil and animals, vegetation and rocks with it, almost denuding the bedrock of its cover—almost, but not entirely. Bog mosses would not be stripped completely, even in the lowlands, while the heathery moorland of the upper slopes might be partially stripped, but not bare. Nor would a single inundation in seawater kill off the remaining plants. The moors are regularly drenched by sea spray, even to the topmost heights of the mountains. (This, incidentally, is true of all maritime communities and is the reason why human coastal populations do not suffer from goiter like many inland peoples; they receive an adequate supply of iodine from the sea spray.) Since the natural island communities, and coastal communities in general, have evolved to withstand hurricanes and sea spray, giant waves and high winds, they will not be completely destroyed by the tsunami.

The next part of this complex event experienced by the Hebridean islands would be a giant dust cloud and acid rain. The bogs and marshes of the islands are acid enough already and lack calcium carbonate, which would prevent too much acidity developing; they are unbuffered, we say. Accordingly, the acidity would increase alarmingly, but since the total heather and bog communities are fully adapted to acid conditions this might not be fatal for whole taxa, only for individuals. The cold and lack of light would be merely one more winter to survive, a worse winter than usual; there would be heavy winter kill, but not complete destruction, of the total community.

The overall picture in the Hebridean islands adds up to one of desolation but not of total habitat destruction or mass extinction. Most individuals of all species would be killed by one mechanism or another, but enough would survive to reestablish the community long afterward. This is a result of the special nature of the community, which has evolved, serendipitously, to withstand most of the happenings we associate with the Cretaceous-Tertiary event. And as with any devastated area today the first plants to recolonize the denuded landscape would be ferns—the fern spike.

Other coastal communities might not fare as well as the heather and bog of granitic islands. Certainly human coastal populations would not. All coastal communities are more or less inured to the effects of salt spray, so along coasts the tsunami, while doing great mechanical damage, would be unlikely to lead to extinction of taxa, at least among the lesser plants; but trees and large mammals are a different matter. The greatest damage might come from acid rain. Plant communities growing on soil that contains some calcium may not withstand acidification as readily as do communities that are already adapted to life without calcium. Unless there is enough calcium in the soil to prevent, in an effective manner, any acidification (a most unlikely situation), a coastal plant community may well be wiped out by acidification alone. Areas such as the shoreline of Chesapeake Bay would suffer a devastating loss of habitat, species, and whole taxa. Indeed, the fauna of the bay itself would likely be wiped out, since it is not adequately protected from acidification—it is not well buffered, even though the community demands some calcium for its survival.

River Deltas

The delta of the Ganges, which forms the largest part of Bangladesh, is perhaps the nearest ecosystem we have nowadays to the deltaic system that supported the dinosaurs of Alberta during the Cretaceous era. For that reason the possible fate of Bangladesh in the face of a cometary impact is perhaps worth considering in some detail.

The initial supernova flash would ignite all vegetation—the forests (such as they are nowadays), the cultivated fields, even the rice paddies. All larger animals (including man) that could not find adequate shelter would be killed at this time. In the succeeding five thousand years before the comets arrive, the delta might well be repopulated from elsewhere with a different flora and fauna—still a flourishing ecosystem, though perhaps with a paucity of species. Then comes the tsunami.

The whole country would be inundated and then acidified. Nothing would be left of the original plant and animal communities. This countryside consists of freshwater and deltaic flats regularly flooded by fresh water; all inhabitants would be quickly killed by an inundation of seawater. The mechanical damage from the tsunami would scour out all the lands of the delta and saturate the surrounding plains with salt. After the retreat of the wall of water, there would be no land left in the delta, no land in the whole of Bangladesh, just seawater.

A Scenario for the Cretaceous-Tertiary Boundary Event

The earth experienced the flash from the supernova 65 million years ago as the first evidence of the horrors to come. For something like 36 hours the flux of visible and near-visible light from the supernova —infrared and ultraviolet as well as visible light—would have been between ten and twenty times that of the midday tropical sun. This would have had very little effect on the marine biota, but on land it would have been enough to start worldwide forest fires. We have evidence of such fires in the soot associated with the Cretaceous-Tertiary Boundary rocks, lying always just under the claystone layer, not so much within it; in other words, the fires came first and the deposit of clay later.

In my opinion these fires alone were enough to destroy the dinosaurs. The only individual land creatures that could have survived were those that could obtain some refuge from the fires, which affected not merely the forests but also the tundra, the reeds growing in the marshes, the open plains, the very mountainsides. Almost all vegetation on the surface of the earth was destroyed by fire. In particular, all trees were destroyed. Equivalents of the modern-day heather community, evolved to withstand repeated fires, would have survived, either by deep roots or else by seeds or spores buried and protected in some way from the fires.

Animals that could not take refuge from the fires would have been killed. Indeed, we have no evidence that any land animal larger than about 25 kilograms survived the Cretaceous-Tertiary Boundary event. Lizards and salamanders, frogs and toads survived because they could hide deep in crevices of the rocks, on land or under water. This does not mean that all individuals survived, but rather that a few could survive the holocaust while the majority were, in fact, slaughtered. But dinosaurs had nowhere to go. They could not hide their bulk in rock crannies like lizards or mice or dive to the depths of a pond like frogs; neither were they evolved to hide in this way at the approach of danger.

Marine life would have been hardly affected by this event. Certainly the intertidal life would have been baked, but this would not lead to extinction of whole taxa, only of individuals.

On land the few survivors would have been very short of food, and many would have starved, completing the destruction of many species. A very few individuals may have been able to survive. The species that would do best would be those that had evolved some kind of dormancy, creatures able to hibernate through a winter or estivate through a drought, creatures that produced resistant spores or seeds to replace adults that had died. An immediate corollary to this is that species in polar climes would have fared better (as indeed we know they did), since these are largely the ones that have such dormancy devices.

Overall, this time of fire may have been devastating to land creatures, but it was unlikely to have in itself caused a mass extinction, except for killing off all the larger land plants and animals, especially the dinosaurs. Most other taxa would have survived in small numbers and quickly recolonized the devastated area.

Many species of arctic birds and mammals share a behavioral feature that may well have contributed to their survival: the habit of burrowing. Both these classes of vertebrates are warm-blooded (or endothermic or homeothermic, in the jargon of biologists). This means that they keep their body temperature at a more-or-less constant level, day and night, even when sleeping. In a temperate climate a warm-blooded animal requires almost ten times as much food as a cold-blooded creature just to manage this feat. For example, a lion must eat its own body weight of meat every eight or nine days, even in the tropics, while the cold-blooded Komodo dragon, a predatory reptile of about the same body weight as a lion, eats its own body weight of meat only every 90 days or so.

One of the strategies warm-blooded animals in cold climates use to conserve heat, and so to lower food requirements, is burrowing. The sleeping chamber of a rabbit's burrow even in winter maintains a steady temperature of about 30 to 34°C. Even we ourselves burrow in a sense: we are only comfortable in bed if the temperature of the air under the blankets is between 30 and 35°C. In my view this burrowing habit of polar mammals and birds contributed to their survival at the Cretaceous-Tertiary Boundary event, sheltering them from the heat of the supernova flash.

Now came a period of recuperation after the flash. Fifty centuries is quite long enough for the land to be recolonized, to establish vegetation and ecosystems not too different from before, except for the lack of dinosaurs. The climate was warm, warmer than today, but some tropical species may have disappeared, while the arctic and polar biota flourished as never before.

Then came the comets, the largest one, or rather the largest piece of one, landing offshore from the Yucatán peninsula, but other fragments of the same celestial body landed in Iowa, China, and Siberia, and other smaller bodies seem to have landed elsewhere within a very short period of time. Either the earth registered impacts from a number of comets, or possibly a single incoming comet broke apart within the inner solar system, the zone of the stony planets, just like comet Biela was observed to do. The largest piece then further split when it entered the atmosphere, making the four pieces that landed in line. And then came catastrophe once again—a catastrophe affecting not only the land creatures this time, but the aquatic organisms as well.

The first destruction of land communities came from the tsunami, the enormous wave thrown up by the Yucatán bolide landing in the sea off the coast. Even on the other side of the world this wave was at least 100 meters high; across the Atlantic it was 250 meters, and nearer home, in North America, even higher. The drawback from this wave denuded the land down to bedrock, killing everything it reached. The deltaic lands of the coast of Alberta, facing onto the interior seaway (Figure 13.1) were sucked out to sea, and in all the coastlands nothing was left alive. Inland everything was drenched in spray.

Over 10^{19} grams of dust from the craters was thrown into the stratosphere, completely cutting off all sunlight from reaching the surface of the earth. Plants died from lack of light, both in the water and on land. Animals died from lack of food. But within about three or four months, enough of this great cloud of dust had settled out, coming down as a rain

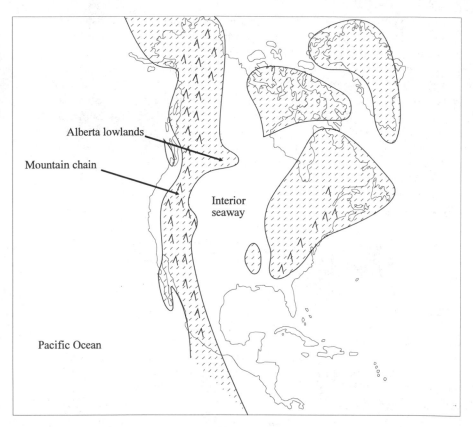

Fig. 13.1. The interior seaway of North America at the end of the Cretaceous. Drawn from data provided by the Geological Survey of Canada.

of mud (including, of course, those extraterrestrial diamonds) covering the whole earth to a depth of a centimeter or more, finally allowing some sunlight to filter through.

Lack of sunlight not only meant a shortage of light for photosynthesis but also a huge drop in temperature, since the heat of the sun had been reflected back to space. Tropical regions were worst affected in this regard, experiencing a temperature drop that may have been as much as 20°C (35°F), killing many species of plants by unaccustomed frosts. And even after enough sunlight began to filter through the thinning cloud of dust to allow some photosynthesis, low temperatures and dim light still persisted for several years.

Once again the polar biota fared better, both because the temperature drop was lower nearer the poles than at the equator, but also because creatures that live in high latitudes almost always have some dormancy device to enable them to survive the winter. Diatoms and dinoflagellates form spores under the polar ice, and caribou moss lives on under the arctic snow. This was for them merely a worse winter—a much worse winter. But the dust cloud was not the only effect of the comets. Far worse was the acid rain produced by their passage through the atmosphere.

I have discussed in Chapter 8 the effects of the passage of a high-speed object through the atmosphere in ionizing the gases of the air and the resulting formation of oxides of nitrogen and thence of nitric acid. A mass of 10^{18} grams entering the atmosphere at over 80 kilometers per second would have produced vast amounts of acid, enough in fact for some rainstorms to have been as acid as battery acid. The long-term effect of these acid rains, continuing for as long as a year, would have been the acidification of all water bodies, including the sea, down to a depth of about 200 meters. The ruination wrought by the stratospheric clouds of dust was completed by this acid rain.

Away from the coasts most of the life in the sea is in the top hundred meters or so of the open ocean, the pelagic zone. Sunlight is able to penetrate in useful quantities (useful, that is, for the life purposes of marine plants) only down to about 200 meters. This top 200 meters is known as the photic zone. Being the zone that is mixed by wave action, it is therefore also known as the mixing zone. We thus have three separate names for this top 200 meters of the open ocean: the mixing zone, the photic zone, and the pelagic zone. Here live almost all the offshore creatures of the ocean: the plants, because they need the light; the animals, because they need the plants for food.

Our computer models of the Cretaceous-Tertiary Boundary event demonstrate that a year after the impact this mixing zone would have been significantly acidified. In the jargon of scientists, the pH of this zone would have dropped by a full unit; in other words, the mixing zone would have become ten times more acid. A drop of one pH unit does not sound like much, but it has enormous effects on the life of the open sea, especially on those creatures, both animal and plant, that possess chalky skeletons. Chalk (calcium carbonate) dissolves freely in acid. Put another way, if the sea is ten times as acid, creatures require ten times as much energy to extract the calcium carbonate from seawater. They simply could not do it. As a result, all the chalk-forming creatures in the pelagic zone were killed, together with the pelagic molluscs, the ammonites, and the belemnites. The formation of the great chalk beds of what is now the North Sea and the white cliffs of Dover ceased abruptly (Figure 13.2).

The young larvae of the ammonites were particularly vulnerable. The adults, with their relatively thick shells, might not have been so much af-

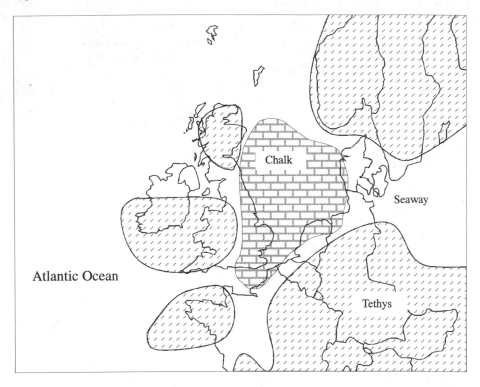

Fig. 13.2. Northwestern Europe at the end of the Cretaceous. Drawn from data provided by the Geological Survey of Canada.

fected by the acid, and in addition some adults may have been inshore, in the neighborhood of coral reefs. But all ammonite and belemnite larvae arc pelagic, living in the plankton, and so were killed by the acid. In contrast, those close relatives of the ammonites, the nautiloids, lay large yolky eggs that develop in the depths of the ocean, taking as long as a year to hatch. Before they penetrate to the mixing zone (which all adult nautiloids do for a few hours every night, sinking down as deep as a thousand meters during the day), they already look like small adults, with fully formed shells. All ammonites and belemnites were killed in the Cretaceous-Tertiary Boundary event; a few nautiloids survived.

Fish were less affected than invertebrates, probably because the mineral in their skeletons is not calcium carbonate but calcium phosphate. Here we have one more example of a feature that evolved for entirely different reasons turning out to be a survival feature during the Cretaceous-Tertiary

Boundary event. Why fish, and vertebrates in general, evolved phosphatic skeletons is still a matter of debate. What is clear, however, is that this is one feature that enables a fish to be much more energetic than a clam. Fast movement, especially if long continued, results in the muscles producing lactic acid. This acid is quite able to dissolve calcium carbonate but not calcium phosphate, as has been shown by implanting rods of these two minerals in the muscles of trout. An ammonite trying to swim fast would have been in danger of dissolving its own shell in the lactic acid produced by hard-working muscles, while a fish does not dissolve the calcium phosphate from its bones under the same circumstances. Thus fish are pre-adapted to withstand acidification of the water in which they live, and so have a better chance of surviving the cataclysm.

Coral reefs too were protected to some extent from the effects of acidification, as were limestone lakes in freshwater zones. The excess lime, whether as limestone or as dead coral, would react with the acid to buffer the water and prevent any significant degree of acidification. This is true to a lesser extent of the continental shelf area in general, so that benthic organisms fared better than the pelagic creatures of the open ocean: seasnails and clams survived (and frogs in limestone lakes and ponds), while ammonites did not. And, of course, creatures that did not rely on calcium skeletons of any kind were less affected. Squid, those shell-less relatives of belemnites and ammonites, survived.

In deposits laid down under the sea the iridium anomaly is found within the claystone layer of the Cretaceous-Tertiary Boundary itself. In the Red Deer River section it lies just above the claystone, in the base of the coal seam. Like most heavy metals, iridium and its compounds migrate into regions of low oxygen, such as would be found in coal-forming conditions. It is my belief that this is what has happened here: the iridium from the comets was attracted to the oxygen-deficient conditions before the rocks were consolidated.

Iodine too is much more concentrated in the coal than in the claystone, and the iodine 129 anomaly is smeared out through the coal layer. The mechanism here is rather different. Iodine is taken up by the roots of land plants, which actively seek iodine. By these two mechanisms the iridium and iodine anomalies cease to coincide with the Cretaceous-Tertiary Boundary layer but are found just above it in the Red Deer section.

As a final note, I must emphasize two points. First, as noted earlier, this scenario applies only to the Cretaceous-Tertiary Boundary and cannot be extrapolated to other mass extinctions, which were almost certainly the result of collisions with extraterrestrial bolides but whose ultimate cause is unlikely to have been the same as this one. We have no evidence to suggest that any other extinction event involved a hypervelocity impactor and hence no need to invoke a supernova. We have no evidence to suggest that any other event involved a comet rather than an asteroid. And we have no evidence to suggest that successive extinction events formed part of a series or had anything to do with one another, apart from all being the result of impacts.

Second, the complexity of the Cretaceous-Tertiary Boundary event, involving both direct effects of the supernova and also both physical and chemical effects of the comets, is the only possible way of offering some kind of coherent account of what survived the event and what was killed off. The diamonds referred to in the title of this book have nothing directly to do with the extinction of the dinosaurs, but they do provide an essential clue in helping us unravel the whole complex event.

Implications for Evolution

When the messenger of death comes to take you away, let him
find you prepared. Alas! You will have no chance to speak, for
truly his terror will be before you.
　　　　—Egyptian inscription from the time of Ramses II

Was death that quick at the time of the
Cretaceous-Tertiary Boundary event? Certainly for those living within the
range of the ejecta blanket—a few hundred kilometers at most from the
site of an impact—death would have been instantaneous. And the mass
extinction occupied a mere moment of geological time. But on time scales
more appropriate to the lifetime of an animal, death was far from instantaneous for most creatures.

I have argued that the dinosaurs were most probably killed by the great
fires ignited by a supernova, before the bombardment by comets. Death
for many of them would have been not from the fires themselves, but
rather from heatstroke or from smoke inhalation. Elephants suffer from
heatstroke, since they have difficulty ridding themselves of excess heat
when they have been forced to overexercise. How much more careful must
the larger dinosaurs have been about cooling off, or rather about never
overheating! It has been argued that the long tails and necks of the sauro-
pod dinosaurs were in part a cooling mechanism. But exposed to the fierce
heat of the exploding supernova, death from heatstroke would have been
fast, though not instantaneous. Some individuals were certainly caught in
the great forest fires and were asphyxiated or incinerated.

Other creatures survived these initial fires only to be exposed to the
second cataclysm of the impacts, with their attendant acid rain. Planktonic foraminifera and coccolithophorids were not killed outright, but
their ability to form calcareous skeletal parts (the material of chalk) was

slowly undermined by the progressive acidification of the pelagic zone of the oceans. Their extinctions may have taken many generations. Adult ammonites may have survived the acidification, but their planktonic larvae could not form shells under the new acid conditions. Once more, extinction presumably took many generations to run to completion. But from the point of view of geological time, and of natural selection, the extinctions of individual species were effectively instantaneous.

Comets as Portents of Doom

The ancient world took it for granted that the apparition of a comet portended great disaster. Indeed, the very word *apparition* suggests evil supernatural forces at work. Our earliest written records of this kind of belief date from ancient Mesopotamia, from the Sumerians and the Babylonians, whose priests interpreted comets as omens of disaster to come.

The Roman empire collapsed in the West around the end of the fourth century, and forty years later, in the same year that Attila the Hun entered Rome, what remained of the British dependency of the empire experienced a major catastrophe marked, according to Spanish astronomers, by a strange comet. The sky over Britain was darkened, and the "ruin of Britain" was accomplished by fire falling from heaven, fire that did not die down until it had burned the whole surface of the land. The commentators of the period are allowed some degree of exaggeration in describing what actually happened, but Chinese astronomers also recorded a major comet at this time.

Up until the seventeenth century danger from comets remained largely the sphere of astrologers and the fear of the common people. Isaac Newton, however, followed by Edmund Halley, put the study of comets on a scientific basis, and suggested that they really did offer a major, though rare, threat to human life. Catastrophism began to have a scientific life of its own.

Cataclysmic Theory

In the last years of the eighteenth century a British engineer, William Smith, while surveying the line of a new canal, noted that the strata of rocks were layered like "so many slices of bread and butter," each with its characteristic fossils. From this observation he began to develop the science of stratigraphy, providing us with many of the names we use today. He named most of the great geological periods and observed that each period ended with a major shift in fauna, with the biggest changes coming at the end of the Permian and of the Cretaceous periods, the uppermost rocks of the Paleozoic and the Mesozoic eras, respectively.

With minor adjustments, Smith's classification of the strata has stood the test of time. The Cambrian period is named for its most common location in Britain—in Cambria, or Scotland. Other periods were named for the ancient British tribes that in the time of Julius Caesar lived in the areas where the rocks are most common—the Silures, the Ordovices. The Devonian, or Old Red Sandstone, he named for the county of Devon in southern England; the Cretaceous, he named after the Latin word for chalk in honor of the white cliffs of Dover. The major coal-bearing strata of England he named the Carboniferous. Some of his names have not survived in our current usage. The New Red Sandstone is now called the Triassic, but Smith's *Geological Map* of Great Britain, published in 1815, remains the foundation of modern stratigraphy, not just in England but throughout the world.

Each of Smith's stadia (meaning stages or periods), he believed, ended in a cataclysm of some kind: Noah's flood perhaps or vulcanism. After all, the excavations of Pompeii and of Herculaneum were well under way at the time he was forming his ideas, and these cities, buried by the eruption of Vesuvius, stood for him as a paradigm of universal destruction in a cataclysm. He recognized, moreover, that many of the beds (rock strata) he saw in canal cuts contained marine fossils; he surmised that they must have been laid down under the sea and later elevated in some kind of cataclysm. Altogether, his observations led him to propose that each geological period ended in some kind of catastrophe, despite all the teachings of the church. And the greatest cataclysms of all were to be seen at the end of the Primary and Secondary eras. Indeed, it was the very size of the faunistic changes at these times that led him to divide the fossil record into Primary, Secondary, and Tertiary, each containing several stadia, all terminating in lesser cataclysms.

Smith's cataclysmic theory—the first scientifically based theory of its kind—never took hold of the minds of geologists of his day, in part because he was not a philosopher, or scientist, as we would say nowadays (to use a word not coined until half a century later). He was instead an engineer who specialized in designing canals, a profession that at that time commanded little respect and less money. Indeed, he was on one occasion confined to a debtors' prison for several weeks. The "natural philosophers" of that period of rigid class distinctions in Britain were almost invariably men of independent means, like Charles Darwin. Some of them held university professorships but did not rely on these as a chief source of income.

Uniformitarianism, Gradualism, and Natural Selection

More acceptable to society at that time were two Scotsmen of independent means, both nominally lawyers but by avocation practicing geologists. The first was James Hutton, who in 1795 published in two volumes a *Theory of the Earth*, which enlarged on a couple of papers he had published earlier. In these publications he presented his theory of "uniformitarianism," maintaining that all features of the earth's crust can be explained by means of natural processes, observable today, acting over a long period of geological time. His ideas were opposed to the current thinking of the time and provoked a stormy controversy of the kind that stimulates scientific advancement. Hutton's ideas left no room for cataclysms.

His compatriot, Sir Charles Lyell, took Hutton's ideas further, and in his influential book *Principles of Geology*, published just before Charles Darwin departed on HMS *Beagle* for his trip around the world, Lyell argued strongly that all geological processes can and must be explained in terms of processes going on today and also that the earth must be at least a million years old. This formed the ultimate academic basis for Darwin's theory of evolution by natural selection. Natural selection can only act on existing characters that have a marginal selective advantage or disadvantage. As Darwin emphasized, it is a slow gradual process, and gradualism became a keynote of his theory of the origin of species by natural selection. "Nature, red in tooth and claw" was no part of his theory, but rather a later poetic accretion and misrepresentation. (Actually, the words were written by Alfred, Lord Tennyson before Darwin wrote his *Origin of Species* but were later applied to his theory of natural selection.) In line with Lyell's concept of constant slight geological change, natural selection was to proceed by constant exposure to selective factors, factors that were constant over generations, though slowly changing over eons.

Catastrophism and Survival

But what happens if creatures are exposed to a sudden violent event that has not been experienced for many, many generations? There is no possibility whatever that creatures can have evolved adaptations specifically to cope with such an event. Natural selection can only deal with local adaptations to changing environments and proceeds stepwise over generations. Long-term evolutionary trends arise by continuous and intense competition in an overcrowded world, a biological arms race, if you like. Plants develop ever more toxic chemicals to defeat the insects that feed on

Fig. 14.1. Argonauta, a close relative of the ammonites that, unlike them, has survived to the present day. Photograph by David Brez Carlisle.

them (not "prey on them," as many writers put it today; a beast of prey is an animal that feeds on other animals, not on plants). Insects develop new means of detoxifying these chemicals, just as they evolve to resist man-made insecticides. Crabs develop ever stronger claws to feed on sea snails, while the snails themselves develop ever thicker and knobbier shells to resist the crabs.

A catastrophe, however, does not just amplify competition, it offers something totally different. At the end of the Cretaceous the magnificently complex and varied ammonites crashed into complete extinction, while their competitors and close relatives, the argonauts and nautiloids, survived in small numbers to our own day as the paper *Argonauta* (Figure 14.1) and the pearly *Nautilus* (Figure 14.2), used as wine cups by medieval popes, mounted in gold by Benvenuto Cellini. The stately and terrible dinosaurs died in the same cataclysm, while the little lizards in their burrows and crocodiles in the water (or at least their eggs buried in the sand) survived. This was not evolution by competition.

At the end of the Permian, in an even more terrible debacle than the one that terminated the Mesozoic era, three of the four major groups of crinoids (or sea lilies) died, leaving only the fourth deep-sea group to survive in small numbers to the present day. At the end of the Ordovician period all families of graptolites except the then-obscure monograptids

Fig. 14.2. Nautilus, another close relative of ammonites that escaped destruction at the Cretaceous-Tertiary Boundary event. Photograph by David Brez Carlisle.

were wiped out; the one surviving family diversified into a huge number of species. There can be no suggestion that the survivors were more competitive in any way than those that were eliminated. Perhaps they were just luckier, or more likely they possessed some feature, evolved for a different purpose, that preadapted them to the conditions of a particular catastrophic event.

We can see hints of this mechanism at work in the marine plant plankton at the time of the Cretaceous-Tertiary Boundary event. The chalk-forming organisms that give their name to the Cretaceous died, leaving the radiolarians, diatoms, and dinoflagellates to flourish. One obvious difference between these groups is that the former have calcium carbonate skeletons, which, as I have argued in Chapter 13, could not have survived the acidification of the ocean that followed the impact; diatoms and radiolarians have an acid-resistant skeleton of silica, and dinoflagellates too lack any chalky skeleton. Moreover, diatoms and dinoflagellates possess resting spores and are particularly adapted to the high arctic, where they may survive for months under the ice in this resting stage. Few coccolithophorids (chalk-forming organisms) are found in the arctic, and those that are have no resting spores. Resting spores enabled the diatoms to survive under the ice through the darkness of winter, and this fortuitous "preadaptation" later ensured their survival through the two-month-long artificial night of darkness and cold caused by the dust cloud of the Cretaceous-Tertiary impact.

Previous evolutionary history has no bearing on ability to survive a catastrophe. Evolution cannot protect against a cometary impact. Natural selection remains a matter of competition operating over generations; a catastrophe actually eliminates competition for a short period and negates natural selection. The extinction of 85 percent of species at the Cretaceous-Tertiary Boundary and the death of more than 99.99 percent of individuals of those few species that survived ensured a dearth of competition. Evolution proceeds by Darwinian natural selection in peaceful periods ("nature, red in tooth and claw?") but not at times of great catastrophes and cataclysms. Perhaps dinosaurs were exterminated simply because they were all large and had no sheltering burrows like lizards or the early mammals. The "terrible lizards" that dominated all other land life in the Cretaceous had no refuge from a cataclysm and left the tiny, terrified, but sheltered mammals to survive and to give rise to mankind.

We might well ask: Are our man-made refuges enough? Are we sufficiently preadapted for all contingencies to enable us to survive the next otherworldly cataclysm? What will happen to us if an asteroid is on a collision course with the earth? Could we deflect it in time? Could we survive as a species if it hit the earth? How many individuals might actually survive such an impact? A few score underground miners perhaps? How big an asteroidal or cometary fragment would be needed to wipe out half the human population? How big (and how fast) to wipe out 99 percent? If the 1908 Tunguska event (caused by an asteroidal fragment a mere hundred meters across, remember) had occurred over Manhattan instead of over an uninhabited area of Siberia, then Manhattan, with all its great buildings, would have been destroyed, and even cities as far away as Montreal would have been severely damaged, with a huge loss of life.

The study of the Cretaceous-Tertiary Boundary event is no mere academic exercise. What we can learn from the extinction of the dinosaurs and other living beings at this epoch and from the study of other catastrophic events may help the human species survive the next thing from outer space, the next cataclysmic tick of the celestial clock.

Reference Matter

Glossary

Achondritic meteorite Lacking chondrules. Most meteorites contain large numbers of near-spherical chondrules; a few do not. Some achondritic meteorites are ejecta from impact craters on the moon or Mars.

Ammonites An order (Ammonoidea) of Paleozoic and Mesozoic pelagic carnivorous mollusks with complex coiled shells, the inner chambers of which served as floats. Some of them were as big as cartwheels. As best we can tell, they produced thousands of tiny planktonic eggs that hatched into planktonic larvae. The name comes from the fancied resemblance of the coiled shell to the horns of Jupiter Ammon, the Roman god in his personification as a ram-headed man, the assimilation of the Egyptian god Amun, or Amen.

AMS Accelerator mass spectrometry. An ultrasensitive analytical procedure for separating and analyzing isotopes. A particle beam of negative ions containing the isotopes for analysis is first "cleaned up" by a low-energy (120,000 volts) mass spectrometer and then injected into a tandem accelerator operating at three million volts. Halfway along this acceleration, the beam passes through an electron stripper. This not merely strips off negative electrons, creating positive ions, but also breaks up any molecular fragments that may be present in the beam, which is then accelerated over a second three million volts before being passed into a high-energy mass spectrometer for final analysis.

Angiosperms Plants that have their seeds enclosed in some kind of covering. This taxon includes the flowering plants and grasses.

Antipodean The point on the surface of the earth that is exactly the opposite of a given point; the other end of a line drawn through the center of the earth.

Aphelion The point in the orbit of a planet, asteroid, or comet where it is farthest from the sun.

Apollo body A celestial body in a highly eccentric orbit that takes it across the orbit of the earth. These bodies appear to be dead comets, both in their orbits (so similar to those of short-period comets) and spectroscopically, showing spectra similar to those of comets, lacking only the volatile materials.

Apparition A comet only becomes visible, even with a telescope, when it enters the inner solar system and is both illuminated and heated by the sun. This is known as an apparition.

Asteroid One of several million astronomical bodies circulating in the solar system between the orbits of Mars and Jupiter. The gravity of Jupiter has prevented them from coalescing to form a planet. They range in size from about one kilometer across to several hundred kilometers. At some stage in their evolution they have been hot enough for the rocks and minerals to have partially melted and for the lighter silicate rocks to have separated from the heavier siderophilic metals. Numerous collisions have split asteroids into fragments, some of which have coalesced again so that composition may range from that of mantle material to that of metallic core material. Carbonaceous asteroids appear to be dead comets captured into the asteroid belt by the gravitational field of Jupiter.

Astronomical unit (a.u.) The average distance between the center of the sun and the center of the earth as it moves in its slightly eccentric orbit around the sun. It is the main unit of measurement in the astronomy of the solar system and is about equal to 150 million kilometers and to about 8 light-minutes.

Bathypelagic An adjective applied to pelagic animals that live below the photic zone, generally below about 100 meters in the sea, often much below. Examples of bathypelagic animals are the pearly *Nautilus* and the coelacanth *Latimeria*.

Belemnites An order (Belemnoidea) of Palaeozoic and Mesozoic pelagic carnivorous mollusks with straight gas-filled shells. As best we can tell, they produced thousands of tiny planktonic eggs that hatched into planktonic larvae.

Benthos Creatures that live on the seabed (or on the bed of lakes and other water bodies). They may be either attached, like corals, or free, like crabs and flatfish. The latter are often said to be demersal, as opposed to pelagic.

Biota The total assemblage of living creatures—bacteria, plants, and animals—in a predefined space or locality. A useful general term.

Bolide Any object entering the atmosphere (from the Greek *bolos*, a "lump," a clod of earth, a blob). This is the general term for meteors, meteorites, and things from outer space, whether they are bits of asteroids or whole comets.

Brachiopods A phylum (Brachiopoda) of shelled animals with a superficial resemblance to clams, though quite unrelated. Once far more abundant than clams, they are now relatively rare.

Carbonaceous meteorite A meteorite containing 5 to 10 percent of carbon, organic compounds, and diamonds. A fragment of a dead comet.

Cartesian space Cartesian coordinates are at right angles to each other, like the axes of a graph. Cartesian space is normal space as we know it intuitively, with length, breadth, and width. It can, however, in some abstract mathematical treatments, assume a higher number of dimensions, each assumed to be at right angles (orthogonal) to all others. Named after the French mathematician René Descartes.

Catastrophism The theory that successive waves of evolution stem from abrupt catastrophes. Originally, Noah's flood was the great catastrophe, in which animals and plants other than those saved by Noah were totally extinguished. Fossils were thought to be antediluvian (i.e., from before the flood). Then came the idea that several catastrophes, with Noah's flood being simply the most recent, totally eliminated those forms of life unsatisfactory in God's eyes, thus providing a new start. More recently, catastrophism has come to mean not the total extinction of all life, but the mass extinction of many species.

Cenozoic era The third era of the Phanerozoic, extending from the end of the Mesozoic era, about 65 million years ago, to the present. It is also called the Tertiary era or, misleadingly, the Age of Mammals.

Cephalopod mollusks A class (Cephalopoda) of free-swimming or mobile mollusks possessing arms or tentacles. The modern members of this taxon include squid, *Octopus*, *Nautilus*, and *Argonauta*. Extinct members include belemnites and ammonites.

Chaos theory A branch of mathematics developed initially to model weather systems (and improve weather forecasting) and now used as a tool in a wider range of science. Using chaos theory it is now possible to model the dynamics of complex events. A chaotic system may behave for a period of time like a stable system, but sooner or later this transient stability breaks down. This is why short-range weather forecasting is feasible but longer-term forecasting is not.

Chondritic meteorite Containing large numbers of near-spherical chondrules. Most meteorites are chondritic.

Chondrules Small near-spherical agglomerations, like frozen droplets, that form the greater part of most meteorites. Some of them may represent the primitive material that first condensed when the solar system was formed.

Chron A complete cycle of the magnetic reversals of the earth, counting backward from the present. The half cycle when the magnetic polarity was opposite that of the present day bears the postscript R (for reversed magnetism). Thus the Cretaceous-Tertiary Boundary lies in the middle of chron 29R, a period of reversed magnetism. The preceding half chron (during the late Cretaceous) was chron 30, and the succeeding half chron (during the early Paleocene) was chron 29. The current half chron is chron 1.

Circadian A circadian rhythm is an inherent rhythm in animals or plants of approximately a day (from *circa diem*, Latin for "about a day"). In most creatures the circadian period is about 25 hours, when they are shielded from such external influences as 24-hour cycles of light or temperature. In normal circumstances the circadian cycle is reset each day by the natural cycles of day and night, a process known as entraining the rhythm to a diurnal cycle.

Class A major taxon of plants or animals comprising a number of orders (e.g., the class Reptilia or the class Mammalia); subordinate to a phylum but superior to an order (see Taxon).

Cleidoic egg An animal egg that possesses a porous shell and that can only complete its development if it is able to lose water by evaporation, thus developing an air space under the shell. All bird and reptile eggs are cleidoic; fish and amphibian eggs are not. Obviously, a cleidoic egg can only develop on land, not in the water.

Coccolithophorids A family (Coccolithophoridae) of single-celled planktonic green algae that armor themselves with the chalky coccoliths.

Coccoliths Complex, platelike, skeletal elements of calcium carbonate formed by single-celled algae. Coccoliths are the main constituent of chalk.

Comet One of trillions of astronomical bodies formed in the outer reaches of the solar system and circulating in the Oort Cloud far beyond the planets, taking as much as a million years to make one revolution. They range from 100 meters to 40 kilometers across, and a few occasionally penetrate into the inner solar system. Only when a comet appears (as an apparition) in the inner solar system does it become visible. Their composition is best regarded as that of a dirty snowball: about 10 percent dirt (much of which consists of organic compounds) and 90 percent "snow" (water ice and methane ice).

Conifers An order (Coniferae) of gymnosperms—rooted plants, mostly trees—that have no flowers but bear their naked seeds in cones. Modern conifers are mainly evergreens with needlelike leaves, but many Cretaceous species had broad leaves, and some early Tertiary species were deciduous.

Copepods An order (Copepoda) of mostly planktonic Crustacea that now dominate the oceans, accounting for as much as 90 percent of the total animal biomass.

Cretaceous period The most recent period of the Mesozoic era, extending from about 144 million to 65 million years ago and terminating in a cataclysmic event that killed off more than half the known species, including the dinosaurs. It is named for the chalk (from the Latin *creta*, for "chalk") laid down in this period.

Cretaceous-Tertiary Boundary The stratum in the geological record that marks the end of the Cretaceous period (and of the Mesozoic era) and the start of the Tertiary era and of the Paleocene period.

Crinoids Sea lilies. A class (Crinoidea) of sedentary marine animals related to starfish and sea urchins that flourished especially in the Paleozoic. Most present-day species inhabit the deep sea.

Cycads An order (Cycadales) of nonflowering land plants that look something like palms, though they are unrelated. They were dominant forest trees in the Mesozoic but were almost wiped out at the Cretaceous-Tertiary Boundary. Today only a few species exist, some of which are used as ornamental bushes in gardens in the southern United States.

Dead comet A comet that has been baked by the sun for long enough to lose all its volatile materials, so that these are evaporated off into space. Apollo bodies appear to be dead comets; carbonaceous asteroids are fragments of dead comets.

Decapods An order (Decapoda) of Crustacea that includes all the largest species—crabs, lobsters, and shrimps.

Demersal An adjective to describe unattached, bottom-living benthos such as fish and cuttlefish.

Dextro Right-handed. Some organic compounds, including amino acids, can exist in two forms that are mirror images of each other: the left-handed, or laevo, form and the right-handed, or dextro, form. Such pairs are known as enantiomers.

Diatoms An order (Diatomacea) of single-celled green algae chiefly inhabiting colder waters. They are characterized by a siliceous skeleton formed of two sections fitting inside one another like an old-fashioned pillbox. Toward autumn they produce resting spores that can survive the long arctic winter darkness.

Dinoflagellates An order (Dinoflagellata) of single-celled algae possessing two flagella, one of which extends out from the body along the line of motion, while the other beats in a groove (the girdle) around the middle, imparting a spinning motion to the organism. They diversified hugely after the Cretaceous-Tertiary Boundary event and are now dominant planktonic forms in temperate waters. Some of them produce toxins and are responsible for the "red tides" that can cause massive fish kills and for paralytic shellfish poisoning. The shellfish themselves are immune to the toxin, which they absorb from the dinoflagellates they feed on, but the stored and concentrated toxin produces paralysis in people and in fish when they, in turn, eat the shellfish.

Dinosaurs An artificial grouping of two orders (Saurischia and Ornithischia) of large reptiles that have little in common except their large size and the era, the Mesozoic, in which they existed. They laid cleidoic eggs (eggs that develop outside the body in air) and did not bear their young alive.

Diurnal A diurnal cycle is one governed by the natural day-and-night cycle of daily changes. This is a slightly different meaning from the more usual idea of "active during the hours of daylight."

Doliolids A planktonic order (Doliolida) of filter-feeding animals distantly related to the vertebrates. These animals are quite transparent and no fossils are known.

Echinoderms A phylum (Echinodermata) of animals that includes the starfish, sea lilies, sea urchins, and sea slugs.

Ecliptic The plane of the ecliptic is the plane in which the planets orbit the sun. It is so named because eclipses of the sun, moon, and other planets can only occur in this plane.

Ejecta The material ejected from an impact crater by the force of the impact. In a smaller crater most of this material forms a curtain that subsides relatively near to the crater. Only in a large high-velocity crater is material thrown up to the stratosphere. Tektites are often a prominent feature of the ejecta curtain.

Empirical adjustment An adjustment made by an engineer in the course of a calculation. To a scientist it is known as a fudge factor.

Enantiomer Many organic compounds, including amino acids, exist in paired mirror images known as enantiomers: the right- and left-handed forms (or in Latin, *dextro* and *laevo*), abbreviated D and L. Such pairs can often turn into one another after a time, a process known as racemization, leading to the formation of racemic mixtures, or mixtures containing roughly equal proportions of the two enantiomers.

Endothermic Warm-blooded; relying on heat produced within the body to keep the body temperature even.

Eocene The second and longest epoch of the Tertiary era.

Escape velocity The speed a rocket must reach to escape from the gravitational pull so that it will not fall back to the surface. Earth's escape velocity is 11.2 kilometers per second; that of the moon is lower.

Euclidean space Normal three-dimensional space, with the three dimensions being at right angles to each other.

Family A taxon of animals or plants typically consisting of a number of genera (see Taxon).

Faunistic changes Changes in the composition of animal communities; the disappearance of some species and appearance of others.

Foraminiferans An order (Foraminifera) of single-celled animals with skeletons of calcium carbonate. They were almost wiped out in the Cretaceous-Tertiary Boundary event but have since diversified. Some of the oozes that form the floor of the shallower parts of the oceanic abyss are formed largely of their skeletons.

Fossiliferous A rock that contains fossils is said to be fossiliferous.

Fossilization The process by which parts of plants, animals, or even bacteria, usually the hard parts, are modified to form a permanent replica. Many processes contribute to fossilization, and most result in a fossil that is a physical image of the original organ, whether it be leaf, bone, or pollen grain. In a relatively few fossils something of the original chemical composition is preserved. This can only occur if the rocks have not been subjected to heating or to undue compression and folding; mountain-building processes always destroy organic components. The rocks near the Cretaceous-Tertiary Boundary in Alberta contain fossils in which much organic material remains.

Free-fall An object in the solar system is said to be in free-fall if it is moving under the influence of gravity alone without any other extraneous source of energy or of momentum.

Fudge factor A factor introduced into a calculation in order to make the answer come out right. An engineer would call it an empirical adjustment.

GCMS Gas chromatography–mass spectrometry; a delicate method of analysis that separates substances in gaseous form by chromatography and then measures the mass of each constituent, leading (usually) to an identification.

Gas giant One of the large, gaseous outer planets—Jupiter, Saturn, Uranus, Neptune—which consist largely of gas (hydrogen and helium), perhaps with a core of metallic hydrogen.

Genus Plural, genera. A taxon consisting of one or more species sufficiently distinct from other groupings to warrant a name of its own. The name of a genus is always a single word, formally a noun, usually written in italics with an initial capital—e.g., *Tyrannosaurus*.

Geocentric A system of measurement taking the earth as the center. The velocity of an impacting object (such as a meteorite) is usually measured relative to the earth, while that of an object moving in the solar system is generally measured relative to the sun.

Gigapascal A measure of pressure. *Giga* means billionfold, so a gigapascal is a billion pascals, or roughly 66 tons per square inch. Named for the French mathematician Blaise Pascal.

Ginkgos A group of gymnospermous trees, mostly extinct but with a single surviving species. They were abundant near the Cretaceous-Tertiary Boundary.

Gradualism The theory, espoused by Darwin, that all evolutionary change has been gradual and that no abrupt evolutionary changes have occurred.

Graptolites An extinct class (Graptolithina) of Paleozoic animals distantly related to the vertebrates. Most of them were planktonic.

Gymnosperms Seed-bearing plants that have naked seeds, unlike the flowering plants. The most familiar examples today are the pines and cedars, conifers that bear their naked seeds in cones.

Hadrosaurs A family (Hadrosauridae) of plant-eating dinosaurs found in the upper Cretaceous.

Heliocentric A system of measurement taking the sun as the center. The earth has a heliocentric velocity that varies according to its position on its elliptical orbit from just over 25 kilometers per second to just over 28 kilometers per second.

Homeothermic Warm-blooded; keeping the body temperature even. Homeo means "approximately equal," as opposed to homo, which means "equal."

HPLC High-precision liquid chromatography; an extremely sensitive technique for chemical analysis.

ICAP Inductively coupled argon plasma analyzer; an ultrasensitive analytical instrument used for measuring low levels of isotopes. Argon gas is maintained as a plasma at a temperature of some 8,000°C or more by the use of inductive coupling, something like a microwave oven but using radio frequencies instead of microwaves. Material for analysis is injected into this plasma and its spectrum measured.

Ichthyosaurs An order (Ichthyosauria) of aquatic Mesozoic reptiles with fishlike bodies and long, toothed beaks. They were carnivorous and their legs had become finlike. They bore their young alive and did not lay eggs.

Impact velocity The speed of impact of a meteorite or bolide.

Iridium anomaly A stratum of rock that contains an anomalously high level of the metal iridium. It was the presence of an iridium anomaly at the Cretaceous-Tertiary Boundary that first led to the hypothesis that this was caused by the impact of an extraterrestrial body.

Isotope All elements exist in several forms, known as isotopes. The number of protons in the atomic nucleus is unique for each element, and any change in this number leads to the formation of a new element. But the same atomic nucleus may have different numbers of neutrons, leading to different isotopes of the same element. Thus carbon, which has six protons, may have six, seven, or eight neutrons. The atomic number of an element is the number of protons; the atomic mass is the sum of the number of protons and neutrons. Thus the atomic number of carbon is 6, while the three isotopes are carbon 12 (the commonest isotope), carbon 13, and carbon 14, with six, seven, and eight neutrons, respectively. If there is too great an imbalance between the number of protons and number of neutrons, the atomic nucleus is unstable and hence radioactive. Thus carbon 14 is a radioactive isotope of carbon, or a radioisotope.

Joule A unit of energy roughly equal to a quarter of a calorie (or to a four-thousandth part of a dietetic calorie, since this is in fact a thousand calories). Named for the eighteenth-century British physicist James Joule.

Jurassic period The middle period of the Mesozoic era, extending from about 208 million to 144 million years ago.

Kirkland gap One of several gaps in the asteroid belt where resonance with the orbit of Jupiter and other gas giants prevents the formation of stable orbits. Named for the nineteenth-century American astronomer Daniel Kirkland, who first observed one of these gaps.

Kuiper Belt A belt of comets in a parking orbit beyond the orbit of Neptune. This belt is unstable and is the source of short-term comets, when their orbits decay. Named for the astronomer Gerard P. Kuiper, who first proved its existence.

Laevo Left-handed. Some organic compounds, including amino acids, can exist in two forms that are mirror images of each other, the left-handed, or laevo, form and the right-handed, or dextro, form. Such pairs are known as enantiomers.

Larvaceans A class (Larvacea) of filter-feeding animals distantly related to the vertebrates. These animals are quite transparent and no fossils of them are known.

Leonids A meteor shower occurring every year in the northern hemisphere on November 17 and radiating from a point in the constellation Leo. This shower is irregular, weak some years and strong others, reaching its brightest every 33 or 34 years. The Leonids are derived from the gradual breakup of a comet and have been observed at least since 902 A.D.

Light-year The distance light moves in one year. The earth is about 8 light-minutes from the sun; the nearest star, about 4 light-years away; and the outer edge of the Oort Cloud of comets, almost 2 light-years.

Littoral zone The zone of the seashore between extreme high- and low-tide marks.

Mafic Dark-colored volcanic rocks that often owe their darkness to their metallic content.

Mass Applied to isotopes, this is simply the sum of the number of neutrons and number of protons in the nucleus. Applied to naturally occurring elements, which may be mixtures of two or more isotopes, it is the mean mass of the natural element divided by the mass of a single proton. This latter usage is not followed in this book, where "atomic mass," or more simply "mass," denotes isotopic mass.

Mesozoic era The middle era of the Phanerozoic, extending from about 245 million to 65 million years ago. It comprises, in succession, the Triassic, Jurassic, and Cretaceous periods and is the era when all the largest animals were reptiles: the dinosaurs on land; the pterodactyls in the air; and the ichthyosaurs, mosasaurs, and plesiosaurs in the water. Sometimes called the Age of Reptiles or the Age of Dinosaurs.

Meteor A small fast-moving body penetrating the atmosphere of the earth from outer space and burning up totally as a shooting star. Most are less than a millimeter across and burn up at a height of more than 90 kilometers, yet are visible to the naked eye on a clear moonless night. Many of them appear to be dust formed from the breakup of comets, following the original orbit of the parent comet and forming meteor showers when the earth, in its annual orbit, passes through this trail of dust. An example is the Perseid meteor shower in August each year. The largest meteor in recorded history seems to have been the Tunguska object, which almost reached the surface and did enormous damage in Siberia in 1908 before finally disintegrating completely a mere 10 kilometers up. Most meteors represent dust fragments, generally no more than a millimeter across, derived from comets.

Meteor shower A meteor shower occurs when the earth passes through the dust trail left behind by a passing comet. Some showers occur regularly on or about the same date, and most of the meteors in the shower appear to come from a single point in the sky, the so-called radiant. An example is the Perseids, which occur every year on August 11 or 12 and radiate from a point in the constellation Perseus.

Meteorite A somewhat larger object than a meteor that enters the atmosphere from outer space and is large enough not to burn up before reaching the surface of the earth. To be classed as a meteorite it must be slowed down by atmospheric drag so that it arrives gently enough not to be destroyed completely on landing. Most meteorites are fragments of asteroids (stony meteorites are fragments of the crust or mantle; iron meteorites, fragments of the core of an asteroid). A few come from the moon or from Mars, and others (carbonaceous meteorites) appear to be fragments of dead comets from which all the ices have been boiled away.

Mixing zone The topmost zone of the sea that is subject to mixing by winds and waves. It generally extends to about 150 to 200 meters' depth, a little deeper than the photic zone.

Monte Carlo method A Monte Carlo computer simulation is a method of modeling natural events using chance within a framework of specific rules. It is a powerful method for investigating just what rules a particular system actually obeys and is named for the famous Monte Carlo casino.

Mosasaurs An order (Mosasauria) of freshwater Mesozoic reptiles of superficially crocodilelike appearance but not related to them. Unlike crocodiles, mosasaurs bore their young alive instead of laying cleidoic (air-breathing) eggs in nests on the river bank.

Nautiloids An order (Nautiloidea) of bathypelagic carnivorous mollusks with coiled shells. Inner compartments of the shell contain gas and serve as a float. Nautiloids, of which five species still exist, lay a few large eggs that are attached to the deep seabed and hatch as nearly fully formed benthic larvae.

Nekton Strong-swimming pelagic animals such as fish and squid.

Nucleosynthesis The process by which heavier elements are formed cosmologically from hydrogen and helium. There are two main processes: one occurs inside stars as they "burn up" their nuclear fuel by atomic fusion; the second, in the explosion of a supernova. The former can only produce elements up to iron of mass 56. All heavier isotopes than this are produced in the second process.

Nutation A slight oscillation of the earth's axis with a periodicity of nineteen years. The nineteen-year cycle of the moon is a result.

Oort Cloud A band of comets surrounding the outer solar system and extending halfway to the nearest star. The cloud contains trillions of comets and was named for the Dutch astronomer Jan van Oort, who first proved its existence.

Ophiucus "The Serpent," a northern constellation.

Order One of the larger taxa of creatures, comprising several families and subordinate to a class (see Taxon).

Ornithischians An order (Ornithischia) of reptiles; one of the two unrelated orders of reptiles colloquially called dinosaurs. The name refers to the hipbones, which are like those of a bird.

Orthogonal Mutually at right angles to one another, like the edges of a brick.

Ostracods An order (Ostracoda) of tiny Crustacea in which the two sides of the chitinous shell have expanded to enclose the rest of the body, thus resembling a miniature clamshell.

Paleocene The first period of the Tertiary era, immediately succeeding the Cretaceous period.

Parsec A unit of astronomical distance roughly equal to 3.26 light-years.

Pelagic An adjective applied to plants and animals that live their lives in water without ever resting on the bottom. The pelagic grouping includes both plankton and also many fast swimmers (the nekton) such as tuna. Most pelagic creatures are confined to the photic zone, the depths to which sunlight can penetrate. If they live below this zone they are said to be bathypelagic.

Perihelion The point in the orbit of a planet, comet, or asteroid when it is nearest the sun.

Periodicity The property of possessing a regular period. The succession of hours is periodic; so is the succession of regular morning coffee breaks, though this may be less regular than the periodicity of hours. A succession of events with no regularity in the time of their occurrence does not show periodicity. Like many statistical happenings there is some degree of subjectivity in deciding whether a particular series does or does not show periodicity.

Perseids A meteor shower, occurring every year on August 11 or 12 and radiating from a point in the constellation Perseus. The meteors in this shower are dust particles derived from the passage of comet 1862 III, also known as P/Swift-Tuttle.

Phanerozoic era The era in which fossils are more or less obvious in sedimentary rocks (from the Greek *Phaneros*, meaning "visible"). It extends from about 590 million years ago to the present and is divided into three eras, the Paleozoic (Primary), the Mesozoic (Secondary), and the Cenozoic (Tertiary).

Photic zone The upper level of the sea and of lakes to which enough sunlight can penetrate to permit photosynthesis in plants. This generally is confined to the top 100 meters of the sea.

Phytoplankton The plant members of the plankton, the main producers of food for all other creatures.

Planet X A hypothetical tenth planet first proposed to account for supposed aberrations in the orbit of Neptune left unexplained by the influence of Pluto. (X is both Latin for "ten" and "the unknown.") It is now known that the residual aberrations in Neptune's orbit are smaller than was once thought and are in fact within the limits of observational error, so that it is premature to propose any such planet. Planet X has been seized on as a possible disrupter of the Oort Cloud, which might lead to comets entering the inner solar system and colliding with the earth.

Planetesimals Small planetlike objects from which the inner stony planets probably formed by conglomeration.

Plankton Pelagic creatures of the sea and fresh water that are at the mercy of current, not being able to swim fast enough to stay in the same place. Many of them are tiny, single-celled creatures, both animals and plants, while others are larger, such as jellyfish.

Plesiosaurs An order (Plesiosauria) of aquatic Mesozoic reptiles with long snakelike flexible necks and legs converted into paddles. They were fish eaters and bore their young alive, not laying eggs.

Precession The slow change in direction of the earth's axis around a circle, with a periodicity of about 25,800 years. The axis maintains its tilt of some 23 degrees, but changes the direction in which it points. One result is the precession of the equinoxes, whereby the vernal equinox, marking the start of the astrological period of Aries, no longer occurs at the moment when the sun "passes into the house of Aries," as it did when it was first defined.

Prograde The orbit of a celestial body, such as a planet, comet, or asteroid, is said to be prograde (or direct) when it is moving around the sun (or, in the case of a satellite, around its planet) in the same direction as the earth. If it is moving in the opposite sense it is said to be in retrograde orbit.

Pterosaurs An order (Pterosauria) of Mesozoic flying reptiles colloquially called pterodactyls. Like birds, but unlike bats, they laid cleidoic eggs (i.e., eggs that can only develop in air).

Punctuated equilibrium The theory, colloquially known among scientists as "punk rock," that evolution occurs stepwise, with little or no change occurring for long periods (equilibrium) followed by relatively rapid change (the punctuation).

Racemic mixture See Racemization.

Racemization Some organic compounds, including amino acids, exist in paired mirror images, the left- and right-handed forms. When left standing, some of these can spontaneously turn into the other form (a process known as racemization), producing a racemic mixture consisting of roughly equal proportions of the two forms. Racemization may take tens of thousands of years to go to completion.

Radiant The point in the sky from which a meteor shower seems to radiate.

Radioisotope A radioactive isotope (see Isotope).

Radiolarians An order (Radiolaria) of single-celled animals possessing an elaborate skeleton of silica. They were almost exterminated at the Cretaceous-Tertiary Boundary but have since diversified again as pelagic herbivores often carrying symbiotic algae.

Radiometric Radiometric dating relies on measuring the ratios of some radioactive isotopes or sometimes on the ratio of a radioisotope to a stable isotope. The most familiar form of radiometric dating is carbon-14 dating, which is limited to the last 60,000 years or there-abouts, but other methods can be used for dating materials as old as the solar system and planets.

Refugium A place of refuge against catastrophe, which may be as small as a burrow or as large as an island or great lake.

Resonance In music, two notes are said to be in resonance when the frequency of one bears a simple arithmetic ratio to the other. The simplest resonance is the octave, when one note has exactly double the frequency of the other. By analogy, in astronomy, two orbits are said to be in resonance when their periodicities bear a simple arithmetical relation to one another. Such resonances induce instabilities in the orbits. The Kirkwood gaps and the gaps in the rings of Saturn are resonance effects, where such resonances have swept certain orbits free by means of these instabilities.

Retrograde The orbit of a celestial body, such as a planet, comet, or asteroid, is said to be retrograde when it is moving around the sun (or, in the case of a satellite, around its planet) in the opposite direction from the earth. If it is moving in the same sense as the earth, it is said to be in a prograde orbit.

Salps A planktonic order (Salpida) of filter-feeding animals distantly related to the vertebrates. Except for the stomach and reproductive organs these animals are quite transparent and may reach quite large sizes. No fossils of any kind are known.

Saltatory evolution Literally, evolution proceeding by jumps. Many species show periods of "stasis" in the geological record when there seems to be no visible change; then, in quite a short period of time, they "jump" to a new form. Also known as punctuated equilibrium.

Saurischians An order (Saurischia) of reptiles; one of the two unrelated orders of reptiles colloquially called dinosaurs.

Scalar A scalar quantity is one that has only size, without any direction. The length of a piece of string is a scalar quantity, and so is the area of a circle.

Scaling condition The condition that something should look more or less the same whatever the scale. An example is a coastline, which is just as irregular on a large-scale map as on a small-scale map and whose headlands are mirrored in the outlines of sand grains on the beach.

Scientific notation In this notation, E represents "exponentiation." In its simplest form, this means that the number that comes after E is the number of zeroes that must be added to the figure before the decimal point. Thus, 1E+03 is another way of writing 1,000, and 2E+03 is 2,000. If the number after E is negative, then it represents the number of zeroes, including the one before the decimal point, that must be placed before the main number; thus, $1E-03$ is another way of writing 0.001.

Secondary era See Mesozoic era.

Siderophilic Literally, "iron-loving." Siderophilic elements are those that tend to be associated with iron in ores and that are more abundant in the cores of planets (and of asteroids) and in comets than in the earth's crust. They include nickel, cobalt, and manganese as well as platinum, gold, silver, rhenium, osmium, and iridium.

Species A natural grouping of individuals sufficiently alike to constitute a breeding population. Biologists have struggled for years to provide a formal definition, and the best still remains that a species is whatever an expert considers it to be. A species name is written as two words, the first being the name of the genus in which it is included, the second word being known as the "trivial name" (e.g., *Tyrannosaurus rex*).

Spherules Miniature spheres. Most tektites are in the form of spherules, as are chondrules in meteorites. They are generally formed from the solidification of droplets of molten rock.

Spinel A hard mineral, consisting of oxides of two metals—in the commonest form, aluminum and magnesium, for which the chemical formula is $MgO \cdot Al_2O_3$. Some forms of spinel are gemstones (e.g., balas ruby, which is colored by a trace of chromium). Chromite, the only ore of chromium, is a spinel in which chromium is substituted for aluminum and iron for magnesium. Some spinels are extraordinarily refractory and resistant to acids.

Sponges A phylum (Porifera) of marine benthic filter-feeding animals possessing a skeleton of calcium carbonate, silica, or organic material (the familiar bath sponge). Before the Cretaceous-Tertiary Boundary event, some of the calcareous sponges formed reefs like coral reefs.

Stochastic A stochastic interval is an interval of time (or of space) whose length bears no relation to adjacent intervals. A stochastic series of intervals may have a certain average but no regular periodicity.

Stony meteorite See Meteorite.

Stony planet One of the four planets of the innermost solar system, Mercury, Venus, Earth, and Mars. For chemical reasons the asteroids may also be considered in this group. All of them show an excess of oxygen over carbon and contain a preponderance of siliceous rocks.

Sublimation Evaporation of volatile materials from the solid state to the gaseous state without ever becoming liquid.

Taxon Any group of animals or plants sufficiently distinctive to have been given a formal taxonomic name. The largest taxa are the kingdoms (such as the animal kingdom); the smallest, a named variety of a farm or garden species (such as MacIntosh apples or Hereford cattle). Examples of the hierarchy of taxa follow:

Kingdom	Animal Kingdom
Phylum	Chordata (the vertebrates and their allies)
Class	Mammalia (generally hairy beasts that suckle their young)
Order	Carnivora (dogs, cats, wolves)
Family	Felidae (great and small cats)
Genus	*Panthera* (great cats)
Species	*Panthera leo* (the lion)
Cultivar	Persian cat

The categories are augmented as necessary by such taxa as superfamily and subclass. Genus and species are generally written in italics; the others are in Roman type, with an initial capital letter. Any of the taxa may be anglicized by simplifying the ending and using a lower-case initial letter.

Tektites Small glassy objects, generally no more than 2 or 3 millimeters across, that are formed from molten rock in an impact crater as part of the ejecta curtain. Many of them solidify as spheres, but some form flattened disks.

Tertiary era See Cenozoic era.

Toich Sudanese word for the flat plains of cotton soil in the southern part of the country. This countryside is impassable during and for long after the rainy season, and the villages are totally isolated.

Triassic The first period of the Mesozoic era, extending from about 245 million to 208 million years ago, when dinosaurs and mammals first appeared in the fossil record.

Tunicates A subphylum (Tunicata or Protochordata) of animals related to and probably ancestral to the vertebrates. The taxon includes ascidians, salps, doliolids, and larvaceans.

Unicorn A mythical animal sometimes used by biologists as a paradigm of an organism that cannot be affected by changes in the world around it. The unicorn mentioned in the biblical reference that heads Chapter 7 is probably a mistranslation of a word that really means wild bull, and "his band" presumably refers to the herd of wild cattle and the impossibility of harnessing them to the plough or the harrow.

Unicorn fallacy A statistical fallacy leading to the appearance of spurious cycles or rhythms where no such cycles are present in the actual data.

Uniformitarianism The theory that all geological events can be interpreted entirely in terms of current processes—that all processes have been essentially uniform throughout time and that no processes have occurred in the past that are not occurring today.

Vector A quantity that has both size and direction, unlike a scalar quantity, which has only size. An example is the length, width, or height of a brick, as taken from a single corner or point.

Vulcanism The total complex of processes, including the invisible subterranean events, that together comprise volcanic activity (from *Vulcan*, the Roman god of fire).

Zooplankton The animal members of the planktonic community.

Readings

Alvarez, Luis W., Walter Alvarez, Frank Asaro, and Helen V. Michel. 1980. "Extraterrestrial Cause for the Cretaceous-Tertiary Extinction." *Science* 208: 1095–1108.

Alvarez, Walter, and Frank Asaro. 1990. "An Extraterrestrial Impact." *Scientific American*, Oct. 1990, pp. 78–84.

Alvarez, Walter, and R. A. Muller. 1984. "Evidence from Crater Ages for Periodic Impacts on the Earth." *Nature* 308: 718–20.

Badziag, P., W. S. Verwoerd, W. P. Ellis, and N. R. Greiner. 1990. "Nanometre-sized Diamonds Are More Stable Than Graphite." *Nature* 343: 244–45.

Bailey, M. E. 1987. "The Formation of Comets in Wind-Driven Shells Around Protostars." *Icarus* 69: 70–82.

Baldwin, B., and Sheafer, Yvonne. 1971. "Ablation and Breakup of Large Meteoroids During Atmospheric Entry." *Journal of Geophysical Research* 76: 4653–68.

Basri, Gibor. 1990. "Solar Systems in the Making." *Nature* 346: 515.

Blake, David F., F. Freund, K. F. M. Krishnan, C. J. Echert, Ruth Schipp, R. J. Lipari, C. J. D. Hetherington, and S. Chang. 1988. "The Nature and Origin of Interstellar Diamond." *Nature* 332: 611–13.

Bruckner, H., J. Maisch, C. Reinecke, and A. Kimonyo. 1991. "Use of α-aminoisobutyric Acid and Isovaline as Marker Amino Acids for the Detection of Fungal Polypeptide Antibiotics. Screening of *Hypocrea*." *Amino Acids* 1: 251–57.

Carlisle, David B. 1992. "Diamonds at the K/T Boundary." *Nature* 357: 119–20.

Carlisle, David B., and Dennis R. Braman. 1991. "Nanometre-size Diamonds in the Cretaceous-Tertiary Boundary Clay of Alberta." *Nature* 352: 708–9.

Chyba, Christopher F., Paul J. Thomas, and Kevin J. Zahnle. 1993. "The 1908 Tunguska Explosion: Atmospheric Disruption of a Stony Asteroid." *Nature* 361: 40–44.

Courtillot, Vincent E. 1990. "A Volcanic Eruption." *Scientific American*, Oct. 1990, pp. 85–92.

Courtillot, Vincent E., and J. Besse. 1987. "Magnetic Field Reversals, Polar Winds and Core-Mantle Coupling." *Science* 237: 1140–47.

Courtillot, Vincent E., G. Féraud, H. Maluski, D. Vandamme, M. G. Moreau, and J. Besse. 1988. "The Deccan Flood Basalts and the Cretaceous-Tertiary Boundary." *Nature* 333: 843–46.

Davis, Marc, Piet Hut, and R. A. Muller. 1984. "Extinction of Species by Periodic Meteor Showers." *Nature* 308: 715–17.

Duncan, Martin, T. Quinn, and S. Tremaine. 1988. "The Origin of Short-Period Comets." *Astrophysical Journal* 328: L69–L73.

Everhart, Edgar. 1982. "Evolution of Long- and Short-Period Comets." In Wilkening 1982, pp. 659–64.

Galimov, E. M. 1991. "Isotope Fractionation Related to Kimberlite Magmatism and Diamond Formation." *Geochimica Cosmochimica Acta* 55: 1697–1708.

Glen, William, ed. 1994. *The Mass-Extinction Debates: How Science Works in a Crisis*. Stanford, Calif.: Stanford University Press.

Gould, Stephen Jay. 1989. "The Wheel of Fortune and the Wedge of Progress." *Natural History*, Mar. 1989, pp. 15–22.

Greenberg, J. M. 1982. "What Are Comets Made Of? A Model Based on Interstellar Dust." In Wilkening 1982, pp. 131–63.

Grieve, Richard A. F. 1990. "Impact Cratering on the Earth." *Scientific American*, Apr. 1990, pp. 66–73.

Halliday, I. 1987. "The Spectra of Meteors from Halley's Comet." *Astronomy and Astrophysics* 187: 921–24.

Halliday, I., A. J. Blackwell, and A. A. Griffith. 1984. "The Frequency of Meteorite Falls on the Earth." *Science* 223: 1405–7.

Head, M. J., and J. H. Wrenn, eds. 1992. *Neogene and Quaternary Dinoflagellate Cysts and Acritarchs*. College Station, Tex.: American Association of Stratigraphic Palynologists Foundation.

Heisler, Julia, and S. Tremaine. 1986. "The Influence of the Galactic Tidal Field on the Oort Comet Cloud." *Icarus* 65: 13–26.

Huss, G. R. 1990. "Ubiquitous Interstellar Diamond and SiC in Primitive Chondrites: Abundances Reflect Metamorphism." *Nature* 347: 159–62.

Laskar, J., and P. Robutel. 1993a. "The Chaotic Obliquity of Planets." *Nature* 361: 608–12.

——— 1993b. "Stabilization of the Earth's Obliquity by the Moon." *Nature* 361: 615–17.

Lewis, Roy S., Edward Anders, and Bruce T. Draine. 1989. "Properties, Detectability and Origin of Interstellar Diamond in Meteorites." *Nature* 339: 117–21.

Lewis, Roy S., Tang Ming, John F. Wacker, Edward Anders, and Eric Steele. 1987. "Interstellar Diamonds in Meteorites." *Nature* 326: 160–62.

Mellosh, H. J., N. M. Schneider, K. J. Zahnle, and D. Latham. 1990. "Ignition of Global Wildfires at the Cretaceous-Tertiary Boundary." *Nature* 343: 251–54.

Muller, R. A. 1988. *Nemesis*. London: Weidenfeld and Nicolson.

Nuth, Joe. 1990. "Diamonds Are for Everywhere." *Nature* 347: 125–26.

Officer, Charles B., Anthony Hallam, Charles L. Drake, and Joseph D. Devine. 1987. "Late Cretaceous and Paroxysmal Cretaceous-Tertiary Extinctions." *Nature* 326: 143–49.

Palmer, Martin. 1991. "Acid Rain at the K/T Boundary." *Nature* 352: 758.

Peale, S. J. 1989. "On the Density of Halley's Comet." *Icarus* 82: 36–49.

Rampino, M. R., and R. B. Stothers. 1984. "Terrestrial Mass Extinctions, Cometary Impacts and the Sun's Motion Perpendicular to the Galactic Plane." *Nature* 308: 709–12.

Raup, David M. 1986. *Nemesis Affair: A Story of the Death of Dinosaurs and the Ways of Science.* New York: W. W. Norton.

Schwartz, Richard D., and P. B. James. 1984. "Periodic Mass Extinctions and the Sun's Oscillations About the Galactic Plane." *Nature* 308: 712–13.

Sears, D. W. 1975. "Temperature Gradients in Meteorites Produced by Heating During Atmospheric Passage." *Modern Geology* 5: 155–64.

Sharpton, V. L., and P. D. Ward, eds. 1991. *Global Catastrophes in Earth History: An Interdisciplinary Conference on Impacts, Volcanism, and Mass Mortality.* Geological Society of America, Special Paper 247.

Silver, Leon T., and Peter H. Schultz, eds. 1982. *Geological Implications of Impacts of Large Asteroids and Comets on the Earth.* Geological Society of America, Special Paper 190.

Stern, S. Alan. 1986. "The Effects of Mechanical Interaction Between the Interstellar Medium and Comets." *Icarus* 68: 276–83.

——— 1988. "Collisions in the Oort Cloud." *Icarus* 73: 499–507.

Stern, S. Alan, and J. M. Shull. 1988. "The Influence of Supernovae and Passing Stars on Comets in the Oort Cloud." *Nature* 332: 407–11.

Tera, F., D. A. Papanastassiou, and G. J. Wasserburg. 1974. "Isotopic Evidence for a Terminal Lunar Cataclysm." *Earth and Planetary Science Letters* 22: 1–21.

Thaddeus, P., and G. A. Chanan. 1985. "Cometary Impacts, Molecular Clouds, and the Motion of the Sun Perpendicular to the Galactic Plane." *Nature* 314: 73–75.

Touma, Jihad, and Jack Wilson. 1993. "The Chaotic Obliquity of Mars." *Nature* 259: 1294–97.

Weissmann, Paul R. 1982a. "Dynamical History of the Oort Cloud." In Wilkening 1982, pp. 637–58.

——— 1982b. "Terrestrial Impact Rates for Long and Short-Period Comets." In Silver and Schultz 1982, pp. 15–24.

——— 1985. "Terrestrial Impactors at Geological Boundary Events: Comets or Asteroids?" *Nature* 314: 517–18.

——— 1990. "The Oort Cloud." *Nature* 344: 825–30.

Wetherill, George W. 1991. "Occurrence of Earth-like Bodies in Planetary Systems." *Science* 253: 535–38.

Wilkening, Laruel L., ed. 1982. *Comets.* Tucson: University of Arizona Press.

Wolbach, Wendy S., I. Gilmour, E. Anders, C. J. Orth, and R. R. Brooks. 1988. "Global Fire at the Cretaceous-Tertiary Boundary." *Nature* 334: 665–69.

Index

In this index "f" after a number indicates a separate reference on the next page, and "ff" indicates separate references on the next two pages. A continuous discussion over two or more pages is indicated by a span of numbers. *Passim* is used for a cluster of references in close but not consecutive sequence.

Library of Congress Cataloging-in-Publication Data

Carlisle, David Brez.

    Dinosaurs, diamonds, and things from outer space : the
great extinction / David Brez Carlisle.

        p.   cm.

    Includes bibliography and index.

    ISBN 0-8047-2392-3 : —ISBN 0-8047-2494-6 (pbk.) :

    1. Extinction (Biology) 2. Cretaceous-Tertiary boundary.
I. Title

QE721.2.E97C37   1995   94-32694

575′.7—dc20   CIP